Study Guide for
FUNDAMENTAL STATISTICS
FOR BEHAVIORAL SCIENCES

Seventh Edition

ROBERT B. McCALL
University of Pittsburgh

with
RICHARD E. SASS

Brooks/Cole Publishing Company

I(T)P® *An International Thomson Publishing Company*

Pacific Grove • Albany • Belmont • Bonn • Boston • Cincinnati • Detroit
Johannesburg • London • Madrid • Melbourne • Mexico City • New York
Paris • Singapore • Tokyo • Toronto • Washington

Sponsoring Editor: *Cynthia Mazow*
Marketing Representative: *Michael Campbell*
Editorial Assistant: *Rita Jaramillo*
Production: *Dorothy Bell*

Cover Design: *Roy R. Neuhaus*
Cover Photo: *J. P. Fruchet/FPG International*
Printing and Binding: *Malloy Lithographing*

For more information, contact:

BROOKS/COLE PUBLISHING COMPANY
511 Forest Lodge Road
Pacific Grove, CA 93950
USA

International Thomson Editores
Seneca 53
Col. Polanco
11560 México, D. F., México

International Thomson Publishing Europe
Berkshire House 168-173
High Holborn
London WC1V 7AA
England

International Thomson Publishing GmbH
Königswinterer Strasse 418
53227 Bonn
Germany

Thomas Nelson Australia
102 Dodds Street
South Melbourne, 3205
Victoria, Australia

International Thomson Publishing Asia
221 Henderson Road
#05-10 Henderson Building
Singapore 0315

Nelson Canada
1120 Birchmount Road
Scarborough, Ontario
Canada M1K 5G4

International Thomson Publishing Japan
Hirakawacho Kyowa Building, 3F
2-2-1 Hirakawacho
Chiyoda-ku, Tokyo 102
Japan

Printed in the United States of America

10   9   8   7   6   5   4   3   2   1

ISBN 0-534-35367-3

CONTENTS

to the student

The Study Guide has been prepared to supplement your text, *Fundamental Statistics for Behavioral Sciences,* Seventh Edition. Each chapter of the *Guide* corresponds to a chapter in the text and consists of six sections: concept goals, guide to major concepts, self-test, exercises, answers, and optional instructions for using three computerized statistical packages.

The concept goals are the focal topics of a chapter, arranged in a simple list; be sure you master these topics. The guide to major concepts is a semi-programmed review of the basic vocabulary, concepts, and computational routines presented in the text. It requires you to complete an integrated series of statements that summarize the main points of the chapter. The self-test measures your mastery of the material from the text and guide, and the exercises offer you an opportunity to apply the methods and principles of statistics to actual numerical data. The optional boxes at the end of the chapter provide instructions on how to use three computerized statistical packages, StataQuest, Minitab, and SPSS, to perform the computations presented in the exercises of the text and *Guide.*

My students have found that following certain procedures seems to make this *Study Guide* more valuable. First, read each text chapter before working on the guide chapter, which is a review of, not an introduction to, the most important material. After you have read the text chapter, read and complete the guide to major concepts. Then return to the text and read it again, filling out your knowledge of the chapter topics. Finally, try the self-test and exercises in the *Study Guide.* The answers to the questions in these sections appear at the end of the *Study Guide* chapter.

I have tried to make the programmed guide to major concepts more interesting by asking you to recall vocabulary and concepts from the text rather than just repeating information from the previous sentence. I have also concentrated on the most important points and have illustrated these with concrete numerical examples. Each paragraph of the programmed review deals with one or two concepts and requires you to complete one or more statements. You will find the correct answer printed in the margin to the right of each blank. As you work through a paragraph, keep the answers covered with a sheet of paper until you have written your own answers for that paragraph. Then pull the paper down far enough to uncover the correct answers. The purpose of the review is to have you recall and use statistical terminology and concepts; you will benefit from this activity only if you recall the material without looking at the answer and only if you actually write the response in the blank.

At various points in the review you will be asked to perform simple calculations or to complete the calculations outlined in an accompanying table. It is sometimes tempting to skip such activities, but many of my students have found that there is a difference between their ability to understand concepts while reading about them, on the one hand, and to apply them in the solution of numerical problems, on the other. Many students have urged me to encourage you to regard the numerical activities and exercises as the primary method of obtaining a thorough understanding of the nature and application of statistics; you should complete them all.

You will also find guided computational examples in the review section. These are tables that outline the computational routine required to calculate a given statistic or to perform a statistical test of significance. Follow these outlines not only for the exercises presented in the *Study Guide* but whenever you need to calculate the statistic in question. If you do so, you will not forget important details or steps and will be less likely to make a simple procedural error.

The self-test questions, although they cannot provide a complete check on your mastery of all the material, do represent a fair selection of basic chapter topics. Thus the self-test functions as a trial examination by which you can evaluate your progress. Since the test covers material from the text and *Study Guide,* you should take it only after you have read the text chapter, done the programmed review, and reread the text to be sure you understand all the important topics in the text in some detail.

The exercises enable you to apply the concepts you have studied to new situations and to make creative use of what you have learned. These exercises will be of greatest benefit to you if you do not look at the answer until you have finished working a problem. Only through doing the exercises can you learn to apply statistical concepts appropriately and accurately to the analysis of data. To make the exercises relatively painless, I have kept the numbers small and the computations as simple as possible. Many exercises present simplified versions of actual experiments that have had a major impact on thought in the social sciences and thus should be especially interesting to you. Although the material presented in this course is simpler than the statistics performed in some of the actual research projects described, the exercises nevertheless demonstrate why a knowledge of statistics is necessary to read, interpret, or perform research in the behavioral sciences.

If you are using one of the three computerized statistical packages, StataQuest, Minitab, or SPSS, you should work the entire programmed unit, including any guided computational examples *by hand* first. Then, before doing the exercises, work through the instructions for the computerized statistical package that you or your instructor has selected. The instructions assume you know Windows and how to access the statistical package on your machine, and they typically show you how to do the guided computational example using the computerized package. Then do the exercises using the program when appropriate. Be warned, however, that the computer will perform the calculations, but it will not help you understand the statistical concepts or interpret the results you obtain. Also, the purpose of learning to use the program is so that you will be able to use it frequently and efficiently in the future—it may actually take you longer to learn and use it to work the exercises for this course than to do them by hand. Be aware, also, that occasionally a computer program will use a slightly different computational method that may produce a slightly different answer than given in the text or guide. These instances are pointed out in the description of the output from each program.

ROBERT B. McCALL

to the instructor

This *Study Guide* has been prepared to aid students in mastering the basic concepts and applications presented in *Fundamental Statistics for Behavioral Sciences*, Seventh Edition. Each chapter of the *Guide* contains six sections, which are described in the paragraphs that follow.

Concept goals. This section is simply a list of important concepts to be mastered; it ensures that the student is aware of the most crucial topics in each chapter.

Guide to major concepts. This is a semi-programmed summary of the major points made in the corresponding chapter of the text. It is intended to be used as a step-by-step review of only the most important topics after the student has read the text. The presentation is stripped of excess explanation, detail, and secondary considerations. Completing this section of the *Study Guide* requires active involvement from the student, who is asked to recall key terms and concepts from the text or to use principles discussed in the text to solve new problems.

The tone of the guide to major concepts is very concrete and applied; it complements the more theoretical presentation of the text. Immediately after each concept is summarized, it is translated into a numerical example that must be solved. I have encouraged students to explore the application of a statistic by making simple computations on different data sets. In Chapter 3, for example, the variance and standard deviation must be calculated for several small distributions that differ in variability. Then the student is asked to plot the distributions and to label them with the values of s^2 and s; thus the graphical presentation of the concept of variability will be associated with the numerical values of the statistics that index it. Similar exercises illustrate (in Chapter 6) how the correlation coefficient varies as a function of the nature of the scatterplot and (in Chapters 10 and 13) how the mean squares in an analysis of variance are influenced by certain treatment effects.

Whenever possible, major statistical routines described in the programmed unit are organized for the student in guided computational examples. These step-by-step outlines clarify the logical and computational details of important techniques. Students who complete these tables by following the instructions provided will be able to use the same format to solve similar problems in the future.

Self-Test. This covers material presented both in the guide and in the text. Questions may require definitions, explanations, computations, or application of principles to new situations. The test can be used by students to evaluate their progress, since all answers are provided at the end of the chapter.

Exercises. For the most part, exercises require the student to apply concepts and computational routines to new sets of data. They may be used as homework or as the basis of class discussion or laboratory sessions. Many of the exercises are based on actual experiments of major interest in a variety of behavioral sciences. All answers are provided at the end of the chapter.

Answers. Answers to the guided computational examples and all exercises are given, including some of the intermediate quantities so that students may locate their specific errors quickly.

Computerized statistical package instructions. Lastly, three sections give instructions, sample input data, and sample output data, typically for the guided computational example, to students who wish to learn to use the StataQuest, Minitab, or SPSS statistical computer packages. This material is intended to be optional. Teachers must provide instructions for gaining access to one of these packages on the particular machines and systems available to their students. Students are assumed to be familiar with Windows operating systems. The directions given in this *Guide* cover only elementary examples of some of the statistics presented in the text; they do not cover all the options and techniques that each package can perform, and they are not a substitute for the package manual.

I would like to express my gratitude to Richard Sass who did most of the work on the computerized statistical package presentations; Mark Appelbaum of the University of California, San Diego, for his statistical and pedagogical advice; and to Catherine Kelley for preparing the manuscript.

Richard Sass thanks Minitab, Inc., for providing copies of the software as part of their Authors Program, Duxbury Press for a copy of StataQuest, Leon Fulcher of Victoria University of Wellington, New Zealand, for the use of SPSS for Windows, and to the New Zealand Council for Educational Research, Wellington, New Zealand, for office space and the use of their computing facilities in the preparation of the computer outputs for the SPSS and StataQuest programs.

ROBERT B. McCALL

PART 1

DESCRIPTIVE STATISTICS

CHAPTER 1

THE STUDY OF STATISTICS

CONCEPT GOALS

Be sure that you thoroughly understand the following concepts and how to use them in statistical applications.

- ♦ Properties of scales: magnitude, equal intervals, absolute zero point
- ♦ Types of scales: ratio, interval, ordinal, nominal
- ♦ Variables vs. constants
- ♦ Continuous vs. discrete variables
- ♦ Real limits
- ♦ Rounding numbers
- ♦ Mathematical operations involving the summation sign, $\sum\limits_{i=1}^{N}$
- ♦ The difference between $\sum X$, $\sum X^2$, and $\left(\sum X\right)^2$

GUIDE TO MAJOR CONCEPTS

The mathematics required to use *Fundamental Statistics for Behavioral Sciences is* quite simple, not more difficult than the level usually required of high school graduates. Indeed, most of the mathematical operations use only elementary algebra. However, if your memory of those operations is a little fuzzy, Appendix I in the textbook and the Appendix at the back of this *Study Guide* will help you review symbols and algebra. If you are not sure whether you already understand this material, try the Self-Test at the end of the *Study Guide* Appendix. If you miss more than two questions, you probably need some review before proceeding.

The study of methods for summarizing and interpreting quantitative information is called _____ .   statistics

There are two types of methods, **descriptive** and **inferential statistics**. Those that organize, summarize, and describe quantitative information are known as _____ _____ , whereas   descriptive statistics
techniques that permit judgments to be made about a larger group of individuals on the basis of data actually collected on a smaller group constitute _____ _____ .   inferential statistics

Statistics are often necessary to interpret measurements because almost all data in the behavioral sciences contain _____ .   variability
Measurements of the same characteristic in a group of subjects can differ in value for three major reasons: **individual differences**, **measurement error**, and **unreliability**. First, the units (often people) being studied are rarely identical to each other, so one source of variability is _____ _____ .   individual differences
Second, behavioral scientists cannot always measure the attribute or behavior under study as accurately as they would like. This type of variability is called _____ _____ .   measurement error
Third, even if a single person is measured twice under the same circumstances, the measured values may not be identical. This type of variability is called _____ .   unreliability

Scales of Measurement

An important part of science is **measurement**, which is the orderly assignment of a value to a characteristic. This quantitative description, or _____ , is accomplished with   measurement

scales of measurement, which are the ordered set of possible numbers that may be obtained by the measurement process. These _____ of measurement may possess properties of   scales
magnitude, **equal intervals**, and an **absolute zero point**. If the measurement of some attribute permits one to say whether that attribute in one case is greater than, less than, or equal to that attribute in another case, then the scale has _____ . If   magnitude
each unit of measurement (for example, one second, one rating point, one degree) represents the same amount of the attribute being measured, regardless of where on the scale the unit falls, the scale possesses _____ _____ . Finally, if there is a   equal intervals
value that represents none of the attribute being measured, then the scale has an _____ _____ _____ .   absolute zero point

     If a measurement scale possesses all three of these properties, it is called a **ratio** scale, because the scale makes it possible to use the ratio between two measurements. For example, if a father is 70 inches tall and his son is 35 inches tall, the father is twice as tall as his son; so height in inches (or centimeters) is measured on a _____ scale. If a scale has magnitude and equal intervals   ratio
but not an absolute zero point, it is called an **interval** scale. Temperature in degrees (Celsius or Fahrenheit) is an example of an _____ scale. Some measurement scales permit only   interval
statements of relative magnitude and are called **ordinal**. Rankings of football teams would be an _____ scale. If a scale does   ordinal
not possess any of these three properties, namely it does not possess
_____ , _____ _____ , or an   magnitude; equal intervals
_____ _____ _____ , it is called a   absolute zero point
nominal scale. Classification of plant species would be a
_____ scale.   nominal
     Thus, if a scale possesses all three properties, it is a
_____ scale; if it has none of them, it is a   ratio
_____ scale; if it has only magnitude, it is an   nominal
_____ scale; and if it has only magnitude and equal   ordinal
intervals, it is an _____ scale.   interval

Variables and Constants

Within any particular context, a **variable** is the general characteristic being measured on a set of people, objects, or events, the members of which may take on different values. If we have a sample of 20 college students and we obtain their Scholastic Assessment Test scores (SAT), then SAT is a _____ that will take on several   variable values within that sample. Even though it has only two values and we usually do not "measure" them, gender may also be a variable if both male and female subjects are in a study. On the other hand, some quantities do not change their value within a given context, and they are called **constants**. The mathematical quantity π always equals 3.1415... and is thus a _____ . Notice that while SAT in the sample mentioned above was a _____ , the average SAT of that sample is a _____ . However, if we obtain several samples of 20 subjects each and calculate the average SAT for each group, then "average SAT" takes on a different value for each sample. In this new context, average SAT is a _____ , but the average taken over all the samples is a _____ for that collection of samples.

constant
variable
constant

variable
constant

Continuous and Discrete Variables

A variable may be either **continuous** or **discrete**. A continuous variable is one that theoretically can assume an infinite number of values between any two points on the measurement scale. The length in centimeters of lines drawn by subjects, the longevity in years of American men, and the time in seconds that a subject delays before responding are all examples of _____ variables, because given any two values, one can always name a possible value that will be between them. Actually, there are an infinite number of potential values between any two points for a continuous variable. In contrast, discrete variables are those for which one can assume only a countable number of values between any two points. Since the number of aggressive acts a preschool child commits must be a whole number (1/2 response is impossible), number of aggressive acts is a _____ variable; there is no possible score value between 5 and 6 acts, for example. Actually, this distinction between _____ and _____ variables is a theoretical one, because in practice we round off measures to some convenient level of accuracy (tenths of seconds, whole seconds, etc.), and from a practical standpoint the rounding makes the measures, though not the variables, discrete.

continuous

discrete

continuous; discrete

Real Limits

Any single value of a continuous variable actually represents a range of values, roughly composed of the theoretically possible numbers that one would round off to the number in question. For example, if one is measuring time in whole seconds, "23 seconds" represents all the theoretically possible values between 22.5 and 23.5 seconds;

similarly, 24 seconds actually stands for values between 23.5 and
24.5 seconds, and so on. The **real limits** of a number, then, are the
points falling one-half measurement unit above and one-half
measurement unit below that number. If a measurement is in whole
seconds, the _____ _____ of 47 seconds are   real limits
_____ and _____ ; if measurement is in tenths of a second,   46.5; 47.5
the real limits of 43.8 seconds are _____ and _____ (the   43.75; 43.85
measurement unit is . 1 second, one-half of which is .05 second); and
if measurement is in hundredths of a second, the real limits of 67.42
seconds are _____ and _____ .   67.415; 67.425

Rounding

If a decimal number is to be rounded to whole units, then the
whole-unit portion of the number is increased by 1 (*rounded up*) if
the fractional value is more than .5; it is left unchanged if the
fractional value is less than .5. Rounding to whole numbers, 34.2
becomes _____ , 25.50001 becomes _____ , and 19.07 becomes   34; 26
_____ . When the fractional portion is exactly .5, an arbitrary   19
convention is adopted: The number is rounded up if the *preceding*
digit is an odd number, but the number is left unchanged if the
preceding digit is an even number. Therefore, 27.5 becomes
_____ (because 7 is an odd number), 28.5 becomes   28
_____(because 8 is an even number), and 33.5 becomes   28
_____ . However, 32.5001 becomes _____ (even though the 2 is   34; 33
even) because .5001 is greater than .5. The arbitrary convention only
applies when the fractional portion is *exactly* .5000... units.
Analogous procedures are used when rounding to tenths, hundredths,
etc.

Summation

A great many calculations in statistics require summing specific
values. Therefore, many formulas employ a shorthand method for
designating the operation of summing. For example, suppose the
scores on variable X for five subjects are as follows:

Subject (i)	X_i
1	3
2	6
3	5
4	9
5	2

The symbol X_i stands for the value of variable X for the ith subject; for the data above, $X_3 =$ _____ , and $X_4 =$ _____ . Frequently, you will need to sum all the X_i scores. For the above example, this operation is symbolized by $\sum_{i=1}^{5} X_i$, in which the Greek letter sigma and its subscripts and superscripts mean "sum the X_i from $i = 1$ to $i = 5$." Written out, $\sum_{i=1}^{5} X_i = X_1 + X_2 + X_3 + X_4 + X_5$. If the number of subjects is N, the operation of summing all N scores for variable X is written $\sum_{i=1}^{N} X_i$, or simply $\sum X$ if it is clear which scores (usually all the available scores) are to be summed. For the data provided above, $\sum_{i=1}^{5} X_i =$ _____ , $\sum_{i=3}^{5} X_i =$ _____ , $\sum_{i=1}^{2} X_i =$ _____ , and $\sum X =$ _____ .

5; 9

25; 16; 9

25

    Since the summation sign is a frequent element in formulas, it is necessary to understand how to perform simple algebra involving it. Consider the expression $\sum_{i=1}^{N} cX_i$ in which c is a constant and X_i is a variable. This expression tells one to sum over all X from i to N the constant c times the variable. In symbols,

$$\sum_{i=1}^{N} cX_i = cX_1 + cX_2 + \ldots + cX_N$$

But, if the constant c is factored out, one has

$$= c(X_1 + X_2 + \ldots + X_N)$$

$$\sum_{i=1}^{N} cX_i = c\sum_{i=1}^{N} X_i$$

This idea is simpler than it looks. Suppose for the data above that the constant $c = 2$, and you want to find the sum of all the scores multiplied by $c = 2$. There are two ways to obtain the required total. First, you could multiply each score by $c = 2$ and then add the five products. Stated in symbols, you could determine ____ .

$\sum_{i=1}^{N} cX_i$

To compute the total, add

$$2\left(\underline{}\right) + 2\left(\underline{}\right) + 2\left(\underline{}\right) + 2\left(\underline{}\right) + 2\left(\underline{}\right) = \underline{} .$$

3; 6; 5; 9; 2; 50

Alternatively, you could add all the scores first and then multiply this sum by $c = 2$. In symbols, obtain _____ by calculating

$$c \sum_{i=1}^{N} X_i$$

$2\left(\underline{} + \underline{} + \underline{} + \underline{} + \underline{} \right) = \underline{}$. Therefore, *the sum of a*

3; 6; 5; 9; 2; 50

constant times a variable equals the constant times the sum of the

variable. In symbols, _____ .

$$\sum_{i=1}^{N} cX_i = c \sum_{i=1}^{N} X_i$$

Consider the expression $\sum_{i=1}^{N} c$ in which c is a constant. This

expression requires adding N c's together. Again, there are two ways

to determine this total. If N is 5 and c is 2, you could add five 2s. In

symbols, _____ is calculated by

$$\sum_{i=1}^{N} c$$

$\underline{} + \underline{} + \underline{} + \underline{} + \underline{} = \underline{}$. Alternatively, since

$2 + 2 + 2 + 2 + 2 = 10$

multiplication is just a short method of adding the same number over

and over again, you could simply multiply five by two. In symbols,

the desired total is _____ , which can be calculated by

Nc

$\underline{}\left(\underline{} \right) = \underline{}$. Therefore, the principle is that *the sum of a*

$5(2) = 10$

constant taken N times equals N times the constant. In symbols,

_____ .

$$\sum_{i=1}^{N} c = Nc$$

Finally, suppose you are asked to find $\sum_{i=1}^{N}(X_i + Y_i)$. This

expression requires the sum of two variables over a group of
subjects. Again, there are two ways to determine the desired total.
You could add the two scores separately for each subject and then
add these individual totals. In symbols, you would determine

_____ . Using the following table of data,

$$\sum_{i=1}^{N}(X_i + Y_i)$$

Subject (i)	X_i	Y_i	$X_i + Y_i$	$X_i Y_i$	
1	2	3	___	___	5; 6
2	7	1	___	___	8; 7
3	5	6	___	___	11; 30
	$\sum_{i=1}^{N} X_i = $ ___	$\sum_{i=1}^{N} Y_i = $ ___	$\sum_{i=1}^{N}(X_i + Y_i) = $ ___	$\sum_{i=1}^{N} X_i Y_i = $ ___	14; 10; 24; 43

you would determine the values for each subject in the $X_i + Y_i$
column and then add those values to obtain the numerical total,
_____ . Alternatively, you could add the X_i for all subjects   24
(total = _____), then add the Y_i for all subjects   14
(total = _____), and then add the two sums together to obtain the   10
numerical total _____ . In symbols, this latter operation is   24

equivalent to _____ . The rule is that *the summation* $\sum_{i=1}^{N} X_i + \sum_{i=1}^{N} Y_i$

of a sum of variables is the sum of the summation of each variable.

In symbols, _____ = _____ + _____ . $\sum_{i=1}^{N}(X_i + Y_i) = \sum_{i=1}^{N} X_i + \sum_{i=1}^{N} Y_i$

It will be important for students to recognize the difference
between adding and multiplying two variables when a summation
across subjects is to be calculated. The difference can be seen by
completing the right-hand column of the above table, labeled X_iY_i.

While $\sum_{i=1}^{N}(X_i + Y_i) =$ ___ , $\sum_{i=1}^{N}(X_iY_i) =$ ___ .   24; 43

Also, it will be necessary for the student to distinguish between
$\sum X^2$ and $\left(\sum X\right)^2$. The sum of the squared X scores is written
symbolically _____ , while the squared sum of the X scores is $\sum X^2$
written _____ . In terms of the data for the three $\left(\sum X\right)^2$
subjects above, $\left(\sum X\right)^2 = \left(\underline{}\right)^2 =$ ___ , while $14^2 = 196$
$\sum X^2 =$ ___ + ___ + ___ = ___ . $2^2 + 7^2 + 5^2 = 78$

SELF-TEST

1. Individual differences, measurement error,
   and unreliability all produce
   a. variability
   b. descriptive statistics
   c. inferential statistics
   d. scales of measurement

2. The failure to obtain the same measured value
   twice on the same person is known as
   a. individual differences
   b. unreliability
   c. measurement error
   d. variability

3. Indicate the type of measurement scale (ratio,
   interval, ordinal, or nominal) for each of the
   following variables.
   _____ a. Age in years of mothers when
   their first child was born
   _____ b. Type of automobile owned
   where 1 = American,
   2 = Japanese, 3 = Other
   _____ c. Army rank where
   1 = Lieutenant, 2 = Captain,
   3 = Major, etc., as a measure of
   authority

_____ d. Ratings from 1 to 10 which define equal steps pertaining to the livability of U.S. cities as judged by survey respondents

4. Do the following measurements assess discrete or continuous variables?
   _____ a. Number of television sets per household
   _____ b. Average take-home pay of a group of employees
   _____ c. Frequency of bar-pressing by an animal in an operant chamber

5. Determine the real limits of the following numbers.
   a. 90   b. 13.7   c. 30.550

6. Round the following to whole numbers,
   a. 4.6   b. 12.5   c. 24.55   d. 9.5

7. Given the following data and $c = 2$, perform the indicated summations.

Subject (i)	X_i	Y_i
1	6	2
2	0	8
3	4	5
4	3	3
5	1	6

a. $\displaystyle\sum_{i=1}^{3} X_i + c$ d. $\displaystyle\sum cX + \sum Y$

b. $\displaystyle\sum_{i=1}^{3} (X_i + c)$ e. $\left(\sum XY\right) + c$

c. $\displaystyle\sum c(X + Y)$ f. $\displaystyle\sum (XY + c)$

EXERCISES

1. Determine the real limits for the following numbers. (Assume that the unit of measurement is the smallest place listed.)
   a. 9.0
   b. 9.000
   c. 77.001
   d. .002

2. Round the following to whole numbers.
   a. 11.5
   b. 22.5
   c. 80.503
   d. 23.4599

3. Given the data at the right and $c = 2$, determine the following values.

Subject (i)	X_i	Y_i
1	2	3
2	4	6
3	7	8
4	1	5
5	3	2

a. $\displaystyle\sum_{i=1}^{4} X_i$ e. $\displaystyle\sum c(X + Y)$

b. $\displaystyle\sum_{i=1}^{5} Y_i$ f. $\displaystyle\sum (cX + Y)$

c. $\displaystyle\sum (X + Y)$ g. $\displaystyle\sum XY + c$

d. $\displaystyle\sum cY$ h. $\displaystyle\sum_{i=1}^{3} (Y_i + c)$

ANSWERS

Self-Test. (1) a. **(2)** b. **(3a)** ratio; **(3b)** nominal; **(3c)** ordinal; **(3d)** interval. **(4a)** discrete; **(4b)** continuous; **(4c)** discrete. **(5a)** 89.5 and 90.5; **(5b)** 13.65 and 13.75; **(5c)** 30.5495 and 30.5505. **(6a)** 5; **(6b)** 12; **(6c)** 25; **(6d)** 10. **(7a)** 12; **(7b)** 16; **(7c)** 76; **(7d)** 52; **(7e)** 49; **(7f)** 57.

Exercises. (1a) 8.95 and 9.05; **(1b)** 8.9995 and 9.0005; **(1c)** 77.0005 and 77.0015; **(1d)** .0015 and .0025. **(2a)** 12; **(2b)** 22; **(2c)** 81; **(2d)** 23. **(3a)** 14; **(3b)** 24; **(3c)** 41; **(3d)** 48; **(3e)** 82; **(3f)** 58; **(3g)** 99; **(3h)** 23.

STATISTICAL PACKAGES

Many chapters in this *Guide* contain a special optional section that provides instructions on how to use three computer statistical packages, StataQuest, Minitab, and SPSS, to make the computations required in some of the guided computational examples and exercises. These packages were chosen because they are readily available through most academic bookstores or computing centers and are commonly used in social and behavioral sciences research settings.

Students should decide with their instructors whether they should use a computer package in conjunction with this course, and if so, they should pick *one* of the packages and use it *throughout* the course (see Foreword to this *Guide*.) Further, the instructions in this *Guide* do not cover all details on the use of Windows, nor do they cover all the features and output each package can deliver. Students should consult the package manuals for additional information. Moreover, the statistical symbols used in the packages will sometimes be different from those used in the text.

The instructions in this *Guide* assume that the package has been properly installed on the hard drive and that you know how to access it. In the instructions that follow, where commands are given in *italics*, it means they appear on the pull-down menus. Instructions in dialog boxes are given in regular type, while information that must be entered from the keyboard appears in **boldface**.

To start a particular program using Windows 3.1, locate the program group and double-click on the program icon. In Windows 95, select the program from the *Start>Programs* taskbar. Windows is a registered trademark of Microsoft Corporation.

StataQuest

Introduction

StataQuest is a statistical package designed for small analyses and developed especially for student use. It is based on the larger product by Stata Corporation of College Station, Texas. StataQuest is distributed by Duxbury Press. The procedures we describe here are based on Release 4.0 for Windows.

Entering data

Start the StataQuest program. Several windows will appear on the screen. The Stata Editor window may be brought up to fill the screen for easier data entry by clicking on the *Editor* button at the top of the screen.

Enter the data in spreadsheet format, with rows as subjects and columns as variables. Press the *Enter* or *Tab* key to register each value as you type it in. Pressing *Enter* takes you to the next cell down; pressing *Tab* goes to the next cell across. You may enter all the values down the first column, and then go on to the next column, or you may enter the data values row by row, filling in each variable across the spreadsheet. The spreadsheet size is limited to 4000 cells.

When you have finished entering all the data you may review the values and correct as necessary by clicking on the cell to be altered and typing in the corrected value. Finally, you must close the Stata Editor by clicking on the *Close* button at the upper right. You may reopen the Stata Editor at any time by again clicking on the *Editor* button.

Inserting a new column or row

To insert a new variable, enter it at the right end of the spreadsheet. You can use the arrow buttons at the top of the screen to change the order of the variables in the spreadsheet. The left arrow (<<) shifts the current variable to be the first variable. The right arrow (>>) shifts the current variable to be the last variable. The maximum number of variables permitted is 25.

Labeling the columns

To label a column with a variable name, double click on the box at the top of the column. A variable definition box will appear in which you may enter a variable name. Enter the new variable name in the box, and press OK. Variable names are limited to eight characters. The variable names that you enter will appear on the output tables and graphs generated.

Saving the data file

To save the data in the spreadsheet for use at a later time, the following command will save the data in a StataQuest data file with the extension *dta*. Go to the pull-down menus at the top of the screen and click on *File*. Move the pointer down to *Save as* and click. This instruction to act on the pull-down menus can be abbreviated as follows:

> *File>Save as*
>> (Type in a name for the data file.)
>> OK

Opening the log file

To save or print a copy of the output that appears in the Results window, it is necessary to open a log file before carrying out an analysis. To open a log file, click on the *Log* button at the top left of the screen, enter a name and drive for the file, and click OK.

Running statistical procedures

To carry out most operations on the data, the *Summaries* or *Statistics* menus will be used. Boxplots and scatterplots appear under the *Graphs* menu. Instructions for these and other statistical operations are given in the remaining chapters of this *Guide*. Textual output from these operations will appear in the Stata Results window and will be written to the log file for printing. Graphic output will appear in a Stata Graph window, and it also may be printed or saved.

Printing the output

To print the output from the log file, use the command:

> *File>Print Log*
>> OK

This command will print all textual output from the session since opening the log file.

Alternatively, you can use the "clipboard" feature of Windows to copy output from the log file and paste it into a word processor of your choice for printing. Click on the *Log* button to bring the log file to the front of the screen, highlight and copy the desired output to the clipboard using the command:

> *Edit>Copy Text*

Then paste it into a word processor running under Windows. To clear old output, click on the *Log* button and close the log file.

To print any graphs generated by StataQuest, click on the *Graph* button to bring the graphic output forward on the screen, then issue the command:

> *File>Print Graph*
> OK

Returning to the Stata Editor

To return to the Stata Editor and perhaps change or add to the data, click on the *Editor* button at the top of the screen.

Going on to a new problem

If you wish to begin working on a new problem, you must clear the spreadsheet. The best way to clear the data and bring up a new tableau is to use the command:

> *File>New*

Likewise, to clear old output so that only new results will be printed, click on the *Log* button and close the log file. Then open a new log file. The Results window can be brought to the front by clicking on the *Results* button, and the Graph window can be brought to the front by clicking on the *Graph* button.

Getting Help

Help with using StataQuest is available through the Help file accessed through the *Help* menu at the top of the screen:

> *Help>Help*

Additional information is available from the *Help* buttons which appear in the various command boxes.

Ending the session

To end the session and exit the StataQuest program:

> *File>Exit*

You will be prompted if there are any files that have not been saved. It is recommended that you do not save the log files unless you wish to print them later, because they are easily recreated by repeating the analysis.

Minitab

Introduction

Minitab is a versatile statistical package with an impressive array of applications. It is available in both Windows and Macintosh formats, and it also comes in a reasonably priced student version. The procedures we describe here are based on Release 11 of the full version of Minitab for Windows. Minitab is a registered trademark of Minitab, Inc. of State College, Pennsylvania.

Entering data

Upon starting the Minitab program, two windows will appear on the screen. The uppermost is the Session window, onto which all textual output will appear. The bottom half of the screen is taken up by the Data window, onto which you may enter the data in standard spreadsheet format, with rows as subjects and columns as variables. To enter data, point and click on the Data window.

You may enter all the values down the first column, and then go on to the next column, or you may enter the data values row by row, filling in each variable as you go down the spreadsheet. When you have finished entering all the data you may review the values and correct as necessary by clicking on the cell to be altered and typing in the corrected value.

Inserting a new column or row

To insert a new variable, click anywhere on the column immediately to the right of the column to be added. Go to the pull-down menus at the top of the screen and click on *Editor*. Move the pointer down to *Insert Columns*, and click. A new blank column will appear on the spreadsheet. This instruction to act on the pull-down menus can be abbreviated as follows:

> *Editor>Insert Columns*

Similarly, to insert a new row in the spreadsheet, point to the case below the row to be added, and carry out the menu command:

> *Editor>Insert Rows*

Labeling the columns

In Minitab, the columns are already designated with default variable names C1, C2, and so on. To label Column 1 with a more meaningful variable name, simply click on the blank space below the column designator C1, type the new variable name in the space, and press *Enter*. Variable names are limited to eight characters. The variable names that you enter will appear on the output tables and graphs generated.

Saving the data file

The information in the Minitab Data window is known as the worksheet. To save the data for use at a later time, the following command will create a worksheet file with the extension *mtw*:

> *File>Save Worksheet As*
>> (Type in a name for the data file.)
>> OK

If you wish to export worksheet data to other applications besides Minitab, the data may be saved as other file types.

To recall worksheet data saved earlier, use the following command:

> *File>Open Worksheet*
> > (Select the saved worksheet file.)
> > OK

Running statistical procedures

To run statistical analyses on the data, the *Stat* menu will normally be used. Boxplots, scatterplots, and other graphical outputs usually appear under the *Graph* menu. Instructions for these and other statistical operations are given in the remaining chapters of this *Guide*. The textual output from these operations will appear in the Session window, and graphic output will appear in separate Graph windows.

Printing the output

To print the results of any Minitab commands that appear in the Session window, use the commands:

> *Window>Session*
> *File>Print Window*
> > Print Range: All
> > OK

This command will print all the output from the session. To print only a portion of the output in the Session window, highlight that portion and choose "Selection" in the Print Range subcommand above.

Alternatively, you can copy output from the Session window and paste it into a word processor for editing and printing. First highlight and then copy the desired output from the Session window using the command:

> *Edit>Copy*

Then paste it into a word processor running under Windows.

Returning to the Data window

After carrying out an analysis which produces output, the Session window will be active, and perhaps fill the screen. To return to the Data window spreadsheet to change or add to the data, issue the menu command:

> *Window>Data*

Going on to a new problem

If you wish to begin working on a new problem, you should clear the spreadsheet. The best way to clear the data and bring up a new tableau is to use the command:

> *File>New Worksheet*

Likewise, to clear old output so that only new results will appear:

> *Window>Session*
> *Edit>Select All*
> *Edit>Delete*

Getting Help

Help with using Minitab is available through Help files accessed through the *Help* menu at the top of the screen. Additional information is available from *Help* buttons which may appear in various command boxes for selected analyses.

Ending the session

To end the session and exit the Minitab program:

> *File>Exit*

You will be prompted if there are any files which have not been saved. You should not save the Session windows unless you wish to print them later, because they can be easily recreated by repeating the analysis.

SPSS

Introduction

SPSS is an extensive statistical program package typically used by many researchers in the social and behavioral sciences. The procedures we describe here are based on Release 7.0 of SPSS for Windows, Standard Version. SPSS is a registered trademark of SPSS, Inc.

Entering data

Upon starting the SPSS program, the SPSS Data Editor will fill the screen. At this point you may enter the data in the spreadsheet format, with rows as subjects and columns as variables. When entering the first data value in a column, there will be a short delay while the SPSS program defines the variable. Pause for a moment to make sure the value you entered appears in the cell before continuing with the second value.

You may enter all the values down the first column, and then go on to the next column, or you may enter the data values row by row, filling in each variable as you go down the spreadsheet. When you have finished entering all the data you may review the values and correct as necessary by clicking on the cell to be altered and typing in the corrected value.

Inserting a new column or row

To insert a new variable that may have been left out, that is, to insert an additional column into the middle of the spreadsheet, click on the column immediately to the right of the column to be added. Then go to the pull-down menus at the top of the screen and click on *Data*. Move the pointer down to *Insert Variable*, and click. A new column will appear on the spreadsheet. This instruction to act on the pull-down menus can be abbreviated as follows:

Data>Insert Variable

Similarly, to insert a new row in the spreadsheet, point to the case below the row to be added, and carry out the menu command:

Data>Insert Case

Labeling the columns

To label a column with a variable name, click on any cell in the column, then issue the menu command:

Data>Define Variable

Enter the variable name in the box, and press OK. SPSS limits variable names to eight characters. The variable names that you enter will then appear on the output tables generated.

Alternatively, it is possible to enter a variable name by placing the pointer in the header area of the column and double clicking. The variable definition box will appear as above.

Saving the data file

To save the data in the spreadsheet for use at a later time, the data may be saved as an SPSS save file:

File>Save as
(Type in a name for the data file.)
Save

Running statistical procedures

To carry out operations on the data, the *Statistics* or *Graphs* menu will probably be the most useful. For example, both the Frequencies and the Crosstabs procedures appear under the *Statistics>Summarize* menu. Boxplots and scatterplots appear under the *Graphs* menu. Instructions for these and other statistical operations are given in the remaining chapters of this *Guide*. The output from these operations will appear in tables or graphs on the screen. They appear as part of the SPSS Output Navigator.

Printing the output

To get a printed copy of the output that appears in the Output Navigator window:

File>Print
Print range: All visible output
OK

This command will print all the output from the session. To clear old output, see the section below on going to a new problem.

Returning to the Data Editor

After carrying out a statistical operation, the SPSS Output Navigator window will fill the screen. To return to the SPSS Data Editor and perhaps change or add to the data:

Window>SPSS Data Editor

Going to a new problem

If you have been using SPSS for another problem, you must close or clear both the data window and the output window. Here is a quick way to bring up a new tableau:

File>New>Data

Likewise, to clear old output so that only new results will be printed:

Window>SPSS Output Navigator
File>Close

Getting Help

Help with SPSS operations and menus is available through the Index of SPSS commands and keywords, accessed through the menu command:

Help>Topics>Index

Additional information is available from the many published books and manuals describing the SPSS program.

Ending the session

To end the session and exit the SPSS program:

File>Exit

You will be prompted if there are any files that have not been saved. You are not advised to save the SPSS Output Navigator files, because these are easily generated if the data files have been saved.

CHAPTER 2

FREQUENCY DISTRIBUTIONS AND GRAPHING

CONCEPT GOALS

Be sure that you thoroughly understand the following concepts and how to use them in statistical applications.

- Frequency distributions, relative frequency distributions, cumulative frequency distributions, cumulative relative frequency distributions
- Class intervals, number of intervals, size of intervals, lowest interval, mid-point of intervals, real and stated limits of intervals
- Histograms and polygons for various kinds of frequency distributions
- Axes, abscissa, ordinate
- Stem-and-leaf displays, depths

GUIDE TO MAJOR CONCEPTS

Frequency Distributions

One of the purposes of statistics is to organize, summarize, and
describe collections of measurements (i.e., data). **Frequency
distributions** help to accomplish this purpose.

There are many types of frequency distributions. The simplest is
a tally of how many times in a sample of measurements each score
value occurs. The score values are listed from the largest at the top
of the table to the smallest at the bottom and then tallies are made
after each score value to indicate how many times that value occurs.
By examining the resulting table, or _____
_____ , one can obtain some idea of the range of score
values present and approximately what score value was typical.

    frequency
distribution

Table 2-1 lists scores on a statistics quiz for a class of 20
students. At the right of the data is a place for you to rewrite all 20
numbers in order of decreasing score value. After you have done
that, fill in the three left-hand columns below — Score Value, with
the highest score at the top; Tally; and f (frequency of occurrence) —
to make a frequency distribution for these data.

Table 2-1 Frequency Distributions for Scores on a Statistics Quiz

Scores				Scores Arranged in Descending Order			
4	6	9	5				
9	8	5	7				
7	10	7	6				
5	3	6	9				
8	7	7	6				

Score Value	Tally	f	Relative Frequency	Cumulative Frequency	Cumulative Relative Frequency
		$N = 20$	1.00		

You can see by simply glancing at the frequency distribution that the most common score was _____ ; the scores ranged between _____ and _____ ; nearly half the subjects scored either _____ or _____ ; and good marks were scores of _____ and _____ , while poor marks were scores of _____ and _____ .

7

3; 10

6; 7

9; 10

3; 4

Additional information can be obtained from other types of frequency distributions, such as **relative frequency**, **cumulative frequency**, or **cumulative relative frequency** distributions. For example, we often want to know what *proportion* of the total number of cases has been assigned each score value, which information is provided by a _____ _____ distribution. It is called *relative* because the frequencies for each score value are given relative to the total number of measurements, that is, relative to N (relative frequency = frequency of occurrence/N). Fill in the Relative Frequency column of Table 2-1. You can now see that a score of 7 was earned by _____ % of the students and that scores of either 6 or 7 were obtained by _____ % of the students.

relative frequency

25

45

It is also useful to know how many or what proportion of scores *were at a given value or below*. If you scored 8 on the quiz, you might be interested in knowing how many students or what proportion of the class had scores of 8 or lower. Distributions that address these questions are known as _____ distributions. In a cumulative frequency distribution, the entry for each score value represents *the number of cases scoring that value or lower*. A cumulative relative frequency distribution is constructed in the same way, but *the entries are proportions rather than frequencies*. The maximum proportion is 1.00, and proportions become percentages when multiplied by 100. Fill in the Cumulative Frequency and Cumulative Relative Frequency columns of Table 2-1 to produce a _____ _____ distribution and a _____ _____ _____ distribution. Notice that the entry for the score of 5, for example, includes people who scored 5 as well as those who scored less than 5. Thus, 5 or lower was scored by _____ % of the students, scores below 5 were recorded for _____ %, scores of 5 or above for _____ %, and scores higher than 5 for _____ %. You can also see that no one scored below _____ , that scores of 6 or lower were recorded for _____ % of the students, and that if you scored an 8, the percentage of students that scored *higher* than you was _____ . See the answer section at the end of this chapter for the completed Table 2-1.

cumulative

cumulative frequency
cumulative relative frequency

25
10; 90
75
3
45
20%

Class Intervals

The distribution above is quite simple in that it includes only a few score values. In larger distributions, **class intervals** containing several score values are sometimes used instead of a list of individual score values. For example, consider the exam scores of 50 statistics students presented in Table 2-2. As before, the first step is to list all the scores in decreasing order. Do this in the bottom section of Table 2-2 according to the scheme provided. Place the highest scores in each column at the top, and write duplicates next to each other in the same box separated by commas. Scores in the 90s have been entered to illustrate the process. Finish the table now.

Check off scores at the top when you write them at the bottom, and count the scores in the bottom to check that you have 50 when you are finished.

The next task is to determine *how many sets* of score values, called _____ _____ , one should have.   class intervals
Unfortunately, there is no simple rule to answer this question. One needs as many as will summarize and accurately portray the data. To get started, Table 2-3 gives the *maximum* number of intervals for a distribution of a particular size N. For example, the distribution of statistics exam scores in Table 2-2 has $N =$___ , and Table 2-3   50
suggests a maximum number of class intervals of ___ .   14

The next question is *how big should the intervals be?* The major requirement is that the set of intervals must include all the score values. Subtracting the smallest score from the largest will tell you the range of score values that the set of intervals must cover. For the data in Table 2-2, the largest minus the smallest score is
_____ – _____ = _____ . If this range is divided by the maximum $98 - 41 = 57$
number of class intervals to be used, one has an estimate of the minimum size the intervals can be and still include all the scores. Since the maximum number of intervals for $N = 50$ was 14, we divide the range of _____ by the maximum number of intervals of   57
_____ to obtain the *minimum* size of interval, which is _____ in   14; 4.07
this case. Since this is the minimum size, we can round this value up to a more convenient size. Usually, we round up to a "round" value, such as 2, 5, or 10 (or some multiple of 10 times these values, like .2, .5, 1.0 or 20, 50, 100). The next "round" values up from 4.07 are
_____ and _____ . The smaller size interval will use a greater   5, 10
number of intervals and give a more detailed picture of the scores; the larger size interval will use a fewer number of intervals and give a more general picture of the distribution. In this case, either choice would be appropriate, and we choose here for illustration the simpler size of 10.

Finally, we need to *determine the lowest interval*. The custom is to start the lowest interval with a value that is equal to or smaller than the smallest observed measurement but also evenly divisible by the

Table 2-2 Statistics Test Scores for Fifty Students

Scores

79	51	67	50	78
62	89	83	73	80
88	48	60	71	79
89	63	55	93✓	71
41	81	46	50	61
59	50	90✓	75	61
75	98✓	53	79	80
70	73	42	72	74
67	73	79	67	85
91✓	67	77	74	77

Ordering of Scores

	90–99	80–89	70–79	60–69	50–59	40–49
–9						
–8	98					
–7						
–6						
–5						
–4						
–3	93					
–2						
–1	91					
–0	90					

Table 2-3 Maximum Number of Class Intervals for Distribution of Size *N*

N	Number of Intervals	*N*	Number of Intervals
10	6	80	18
15	8	90	19
20	9	100	20
25	10	125	21
30	11	150	22
40	13	200	23
50	14	350	25
60	15	500	27
70	17	1000	30

size of the interval. Since the smallest score in our data is ___ and   41
the interval size is ___ , we choose a number equal to or smaller than   10
___ that is also evenly divisible by ___ . This number is ___ .   41; 10; 40
    The first interval starts with ___ , but what does it end with? We   40
have decided that the size of the interval will be 10, and therefore the
interval must contain 10 possible score values. Beginning with 40,
the 10 score values would begin with ___ and end with ___ . These   40; 49
are called **stated limits**. Since the first interval will include the scores
___ through ___ and the second interval will include the scores   40; 49
___ through ___ , 40-49 and 50-59 are called the _____   50; 59; stated
_____ of these class intervals.   limits
    It is important at this point to recall the concept of **real limits**. If
the measurement is in whole numbers, then the real limits of 40 are
____ and ____ , and the _____ _____ of 49 are ____ and   39.5; 40.5; real limits; 48.5
____ . Since 40 and 49 are the stated boundaries of the class interval,   49.5
it follows that the real limits of that interval are ___ and ____ . The   39.5; 49.5
midpoint of this interval is the number halfway between these real
limits, which is ____ .   44.5
    Now, in Table 2-4, fill in the class intervals for the data in
Table 2-2 beginning with the highest at the top, and then complete
the remainder of the table.

Table 2-4 Frequency Distributions for the Fifty Statistics Test Scores in Table 2-2

Class Interval	Real Limits	Midpoint	f	Relative Frequency	Cumulative Frequency	Cumulative Relative Frequency

A summary of the steps just described to be used in constructing distributions with grouped data is presented in Table 2-5. For additional practice, consider the hypothetical data presented in Table 2-6. A fundamental concept in the study of achievement is *locus of control*, the extent to which persons believe they personally produce, influence, or control their lives and what happens to them. People who strongly believe that they control the events of their lives are said to be *internally controlled*, while persons who think "things just happen to them" or occur because of luck are said to be *externally controlled*. As might be expected, internally controlled people are more motivated to achieve — to try to attain standards or obtain benefits for themselves — because they believe they largely control these events. Conversely, externally controlled individuals are less motivated, because they believe their efforts are largely irrelevant to what happens to them. A test of locus of control is available, and the scores of 50 individuals are presented in Table 2-6, with higher scores associated with internal control. In the spaces provided, arrange the scores in descending order of magnitude, determine an appropriate set of class intervals, and in Table 2-7 construct the four distributions requested.

Table 2-5 Steps in Constructing Frequency Distributions with Grouped Data

1. Obtain an estimate of the **maximum number of class intervals** from Table 2-3.
2. Calculate the **range** of score values by subtracting the smallest score value from the largest score value in the distribution.
3. Obtain an estimate of the **smallest size of class interval** by dividing the range (obtained in Step 2) by the maximum number of class intervals (obtained in Step 1).
4. **Round up** the estimate of the smallest interval size (obtained in Step 3) to the next "round number" (i.e., round up to .1, .2, .5, 1, 2, 5, 10, 20, 50, 100, etc.).
5. Determine the **lowest class interval** so that its lowest *stated* limit is evenly divisible by the size of the interval (obtained in Step 4).
6. Place the lowest interval at the bottom of the frequency distribution.
7. After constructing the distribution, determine whether the distribution accurately describes the data, and adjust the size and number of intervals if appropriate.

Table 2-6 Locus of Control Scores for Fifty Individuals

Scores

28	16	21	13	25	16	27	17	25	20
21	22	29	23	20	21	19	21	20	22
26	14	27	15	26	29	24	11	28	18
19	25	17	27	26	16	19	26	16	21
26	20	23	20	21	22	25	13	23	22

Ordering of Scores

Table 2-7 Frequency Distributions for the Locus of Control Scores in Table 2-6

Class Interval	Real Limits	Midpoint	f	Relative Frequency	Cumulative Frequency	Cumulative Relative Frequency

(Answers will depend upon the specific class intervals used. Find a classmate who used the same intervals that you did and compare answers.)

Graphing

At least two kinds of graphs may be drawn for any of the frequency distributions described above, a **histogram** and a **polygon**. One is composed of vertical bars and is called a frequency

_____ . The other shows frequency points connected by   histogram
straight lines and is known as a frequency _____ . The   polygon
steps in constructing these graphs are described below and
summarized in Table 2-8.

We will use the frequency distribution for statistics test scores
that you obtained in Table 2-4 to construct a frequency histogram in
Figure 2-1. You are given only the **axes**; the **abscissa** and the
ordinate. The horizontal line, or _____ , should be   abscissa
marked off in _____ _____ . Starting from the   score values
left with the midpoint of the lowest interval that you derived in
Table 2-4, number the tick marks on this axis and label the axis. The
leftmost value marked off on the horizontal _____ , or   axis
_____ , will be _____ , and the rightmost value will   abscissa; 44.5
be _____ . The axis will be labeled as _____   94.5; Statistics
_____ _____ . The vertical axis is called the _____ ,   Test Score; ordinate
and it is marked off in _____ . Beginning with 0 at the   frequencies
bottom, number the tick marks and label this axis.

The histogram is constructed by erecting a bar between the real
limits of each class interval, rising to a height equal to the frequency
of that interval and having a width equal to the width of the class
interval. The leftmost bar of this frequency _____ would thus   histogram
rest on the abscissa between the real limits of the lowest class
interval, which are _____ and _____ . It would rise to a   39.5; 49.5
height of _____ frequency units. The next bar would span the   4
abscissa from _____ to _____ and be   49.5; 59.5
_____ frequency units tall. Complete this frequency histogram,   7
and then construct one in Figure 2-2 for the locus of control scores
that you described with a frequency distribution in Table 2-7.

Table 2-8 Steps in Constructing Histograms and Polygons

Frequency Histogram
1. Draw two axes, with the ordinate approximately three-fourths as long as the abscissa. Mark off the abscissa with values corresponding to the midpoints of the class intervals, and mark off the ordinate in frequencies. Label the axes appropriately.
2. Construct a bar of the histogram over each score value or class interval so that the width of the bar covers the score value or class interval from its lower to its upper real limits (not from midpoint to midpoint) and the height corresponds to the frequency of scores in the interval. There should be no space between bars (except if the abscissa is a nominal scale.)

Frequency Polygon
1. Mark off and label axes as for a frequency histogram, but add one interval below the lowest and one above the highest class interval; assign them 0 frequencies.
2. Place points corresponding to the frequencies of each interval (including the two 0-frequency intervals) directly over the midpoints of each class interval. Connect all adjacent points (including the 0s) with straight lines.

Relative Frequency Histogram or Polygon
1. Plot these in the same way as above, but label the ordinate (and locate the heights of the bars or points) to correspond to relative frequency, not frequency.

Cumulative Frequency Histogram or Polygon
1. Follow the steps for constructing a frequency histogram or polygon, except:
    a. Mark off and label the ordinate for cumulative frequency rather than for frequency.
    b. In drawing a cumulative frequency polygon, place the points over the upper real limit of each class interval, including the lowest interval of 0 accumulated frequencies (note that there is no upper 0-frequency interval).

Cumulative Relative Frequency Histogram or Polygon
1. Plot these in the same way as above, except label the ordinate (and locate the heights of the bars or points) to correspond to cumulative relative frequency.

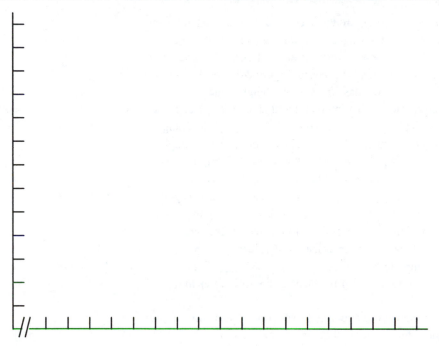

Fig. 2-1. Frequency histogram for the statistics test frequency distribution in Table 2-4.

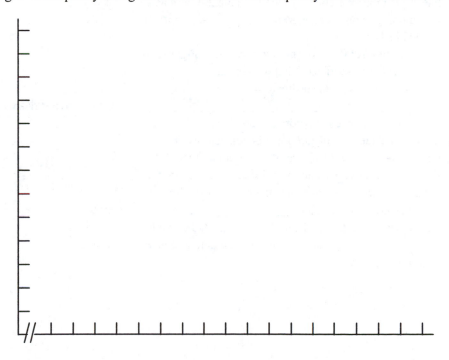

Fig. 2-2. Frequency histogram for the locus of control frequency distribution in Table 2-7.

For **a frequency polygon**, the axes should be labeled in the same way as those for the frequency histogram, except that the abscissa should include one additional interval on the left and one on the right. Mark off and label the axes in Figure 2-3. Locate points over the midpoint of each class interval at a height equal to the number of frequencies in that interval. Do this now. The last point should have been placed over the score value of _____ at a height of _____ frequency units. Connect adjacent points with straight lines. To make this a polygon (i.e., a closed linear form), the first and last points must be connected to the abscissa at locations corresponding to the _____ of the empty intervals you added just below and just above the intervals containing the lowest and highest scores in the distribution. In this case these zero frequencies would be placed over the score values of _____ and _____ . Complete the frequency _____ , and construct one in Figure 2-4 for the locus of control scores in Table 2-7.

94.5

4

midpoints

34.5

104.5; polygon

To graph a **relative frequency histogram** or **polygon**, simply follow the rules above, except that the vertical axis, or _____ , is marked off in units of relative frequency (proportions). Plot the relative frequency data in Table 2-4 both as a histogram and as a polygon on a single graph in Figure 2-5, using the guidelines provided in Table 2-8.

ordinate

To plot a **cumulative relative frequency polygon** in Figure 2-6 for the data in Table 2-4, mark off and label the abscissa as before, and mark off and label the ordinate with _____ _____ _____ . In contrast to a relative frequency polygon, the points in a cumulative relative frequency polygon must be located over the *upper real limits of the class intervals*, not over the midpoints of the intervals. This is done because each point indicates the proportion of scores *within and below* that interval. Thus, the point for the first interval would be placed over the value _____ . Also, a cumulative relative frequency polygon should be connected at the left with the _____ . Complete the plot of this cumulative relative frequency polygon in Figure 2-6.

Cumulative Relative Frequency

49.5

abscissa

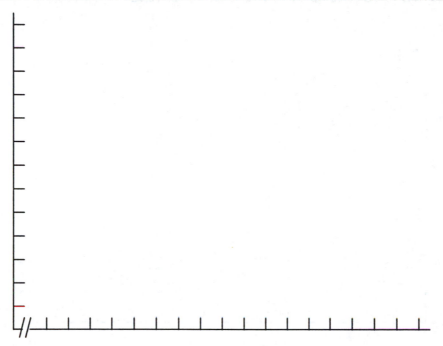

Fig. 2-3. Frequency polygon for the statistics test frequency distribution in Table 2-4.

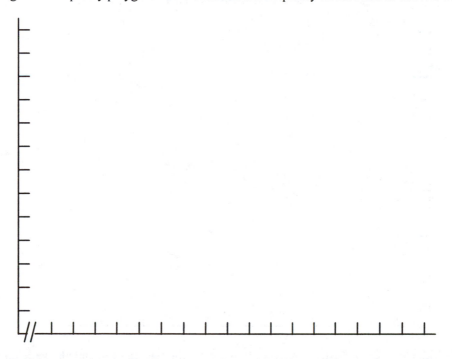

Fig. 2-4. Frequency polygon for the locus of control frequency distribution in Table 2-7.

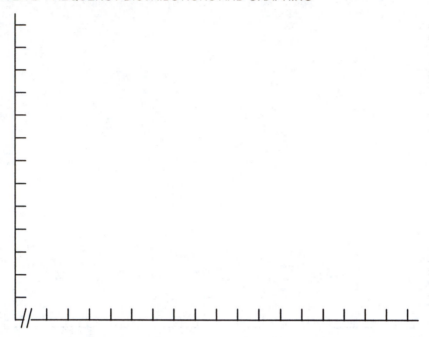

Fig. 2-5. Relative frequency histogram and polygon for the statistics test relative frequency distribution in Table 2-4.

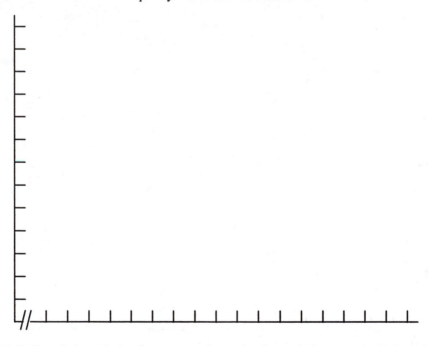

Fig. 2-6. Cumulative relative frequency polygon for the statistics test cumulative relative frequency distribution in Table 2-4.

To review: For a frequency histogram, the horizontal axis, known as the _____ , is marked off in the units of the measurement scale with points located at the _____ of the class intervals. The vertical axis, known as the _____ , is marked off in _____ . The axes should be appropriately _____ . Each bar of the histogram should rise to a height equivalent to the _____ of scores in that interval, and the sides of the bar should be located at the _____ _____ of the class interval. A frequency polygon has the same axes as the histogram, but the points representing frequency are placed over the _____ of the intervals. The line connecting the points in the polygon meets the abscissa on both left and right, at the _____ of the next lowest and next highest intervals. For a relative frequency histogram or polygon, the axes are the same as for the frequency histogram and polygon, but the ordinate is marked off in units of _____ _____ . Cumulative graphs require that the ordinate be marked off in _____ units. For a polygon showing cumulative frequency, the points are placed over the _____ _____ _____ of each interval. The left end of the line connecting the points meets the _____ at the midpoint of the next lowest interval (containing no scores). These guidelines are summarized in Table 2-8 for your convenience in working the exercises.

	abscissa
	abscissa
	midpoints
	ordinate
	frequencies
	labeled
	frequency; real limits
	midpoints
	midpoints
	relative
	frequency
	cumulative
	upper real limit
	abscissa

Stem-and-Leaf Displays

The distributions and graphs presented above are the traditional methods of describing a set of data. Newer approaches, called **exploratory data analysis**, have offered the **stem-and-leaf display** as an alternative to the _____ _____ and _____ for grouped data.

For the purpose of constructing a _____-_____-_____ _____ , the numbers in the distribution, which will now be called a **batch**, must be broken down into a part that consists of the smallest digit, called the _____ , and the remaining larger digits, called the _____ . For example, the number 57 would have a stem of _____ and a leaf of _____ , and the number 23.4 would have 23 as its _____ and .4 as its _____ .

	frequency distribution
	histogram
	stem-and-leaf
	display
	leaf
	stem
	5; 7
	stem; leaf

Similar to a frequency distribution, a ____-_____-_____   **stem-and-leaf**
_____ is an ordered set of all the numbers in the _____ in   **display; batch**
which the class intervals are defined by the _____ and the size of   **stems**
the class interval is defined by the number of _____ within a stem.   **leaves**

The steps in constructing a stem-and-leaf display, a technique of the general approach to descriptive statistics known as

_____ _____ _____ , generally   **exploratory data analysis**
follow those given in Table 2-5 for frequency distributions, but with a few modifications. As before, start by ordering the data by score value, beginning with the smallest number at the bottom. You did this in Table 2-2 for the $N = 50$ statistics test scores.

Second, determine generally the stem and leaves for the data. The data in Table 2-5 range from 41 to 98, so the tens digit will be the _____ and the ones digit will be the _____.   **stems; leaves**

Third, following the procedures described above, determine the approximate number of class intervals, which will be called **lines** in the context of a ____-_____-_____ _____ . Recall   **stem-and-leaf display**
from constructing the frequency distribution for these data that Table 2-3 suggested that the maximum number of intervals for $N =$ _____ was _____ . Further, the range of score   **50; 14**
values for these data was equal to _____ – _____ = _____.   **$98 - 41 = 57$**
So, as before, the minimum size of an interval is $57/14 =$ ___ .   **4.07**
Then, following the steps in Table 2-5, this minimum size of interval should be rounded *up* to the next "round" value, that is either 2, 5, or 10 (or some multiple of 10 times these "round" values, such as .2, .5, 1.0 or 20, 50, 100). So the interval size, which is called the **line width**, for this example will be either ___ or ___ . Recall, for   **5; 10**
simplicity, we chose an interval size of 10.

Fourth, identify the lowest score, which is ___ , and determine   **41**
the lowest line that would include that score. The score of 41 has a stem of ___ , and since the _____ _____ is 10, the stem of 4   **4; line width**
will have ___ leaves beginning with 0. In Table 2-9, place at the   **10**
bottom the first ____ of this display, having a stem of _____ .   **line; 4**
Now add all the remaining possible stems until you reach at the top of the display a stem that includes the largest score, ____ , which   **98**
stem is _____ .   **9**

Fifth, go to the ordered data in Table 2-2 and place the leaves of each score on the line corresponding to its stem, beginning with the smallest leaf at the left and increasing in value to the right along the line. These leaves are separated only by a space (i.e., no commas). If no scores exist for a possible stem, keep the stem in the display but enter no leaves.

Sixth, at the left of Table 2-9 you will find a column labeled **depths**. The cumulative frequency of each line is roughly that line's _____ . Calculate the depth of each line, beginning at the bottom of Table 2-9, until you reach the line containing the middle score (between the 25th and 26th score for $N = 50$). That line gets a depth equal to the frequency, not cumulative frequency, of that line alone. This value is ____ , and it is enclosed in parentheses to indicate that it is not a cumulative value. Then start at the top of the display, and accumulate frequencies for each stem, or _____ , downward until you again reach the line containing the middle score. Note that if a line has a depth that exactly equals half the number of cases in the _____ (25 in this case), there will be two adjacent lines with that depth and no central line with only its frequency.

depth

18

line

batch

Seventh, label the display. Specifically, give it a title (as already printed for Table 2-9), place the N under the heading Depths, and write the unit of measurement under the title. The unit in this case is _____ point. The completed Table 2-9 represents a _____-_____-_____ _____ .

1

stem-and-leaf display

The above example used a size of class interval or _____ _____of 10, so the stems were 10's and the leaves for each stem were the digits 0-9. But we could have used a line width of 5 or, if N were larger, even 2. In this event, a single stem must be divided into two (or five) to produce line widths of 5 (or 2). For example, if the line width were 5, the stem of 4 would be represented twice, once for the five leaves 0-4 and once for the five leaves 5-9. To symbolize this, the stem 4* represents the stem of 4 having leaves 0-4, and the stem 4· represents the stem of 4 having leaves 5-9. In such a display, a score of 53 would be listed next to the stem of _____ , and the score of 77 would be listed next to the stem of _____ . If 2 rather than 5 is the _____ _____ , then there will be five lines for each stem. In this case, a stem plus

line width

5*

7·

line width

* designates leaves 0 and 1
t designates leaves 2 (two) and 3 (three)
f designates leaves 4 (four) and 5 (five)
s designates leaves 6 (six) and 7 (seven)
· designates leaves 8 and 9.

In this event, a score of 35 will have a stem of _____ , 48 a stem of _____ , 61 a stem of _____ , 96 a stem of _____ , and 73 a stem of _____ .

3f

4·; 6*, 9s

7t

Table 2-9 A Stem-and-Leaf Display for the Statistics Test Scores in Table 2-2

Statistics Test Scores

Depths		(Unit = 1 _____)	
(N =)		Stems Leaves	

SELF-TEST

1. What is the purpose of making a frequency distribution?

2. Define:
    a. frequency distribution
    b. relative frequency distribution
    c. cumulative relative frequency distribution
    d. class interval

3. Describe how to determine the following:
    a. the number of class intervals
    b. the size of the class intervals
    c. the lowest interval

4. If 90 and 95 are the stated limits for a class interval,
    a. what are the real limits of this interval?
    b. what is the size of the interval?
    c. what is the midpoint of the interval?

5. True or false?
    a. ___ The vertical axis is the ordinate.
    b. ___ In a frequency polygon, the points are placed over the lower real limits of each interval.
    c. ___ In a frequency polygon, the first and last points always show frequencies of 0.

    d. ___ In a frequency histogram, the bars span from midpoint to midpoint of adjacent class intervals.
    e. ___ In a cumulative frequency histogram, the bars span from the lower to the upper real limits of each interval.
    f. ___ In a cumulative relative frequency polygon, the points are always over the midpoints of the intervals.
    g. ___ In cumulative polygons, the height of the first point is always 0.

*6. When should a histogram and when should a polygon be used to display data?

*7. Characterize in words and with graphs:
    a. central tendency
    b. variability
    c. skewness
    d. kurtosis

*8. If the skewness increases or a distribution becomes more platykurtic, what will happen to the variability of the distribution?

9. A stem-and-leaf display is an alternative to
    a. a frequency distribution
    b. a relative cumulative frequency distribution

c.   a histogram
d.   both a frequency distribution and
     histogram

10. What are the similarities and differences
     between cumulative frequency and depths?

Questions preceded by an asterisk can be answered on the basis of the discussion in the text, but the discussion in this Study Guide does not answer them.

EXERCISES

1.   Below are the high temperatures in Fort Lauderdale for the first 25 days of April. Without grouping the data into class intervals, make a frequency distribution, relative frequency distribution, cumulative frequency distribution, and cumulative relative frequency distribution. Then, separately for each of these four distributions, draw a histogram and a polygon.

Temperatures

80	78	75	75	74
75	73	71	73	74
70	79	76	76	77
73	71	77	77	76
77	74	75	76	76

Score Value	f	Relative Frequency	Cumulative Frequency	Cumulative Relative Frequency

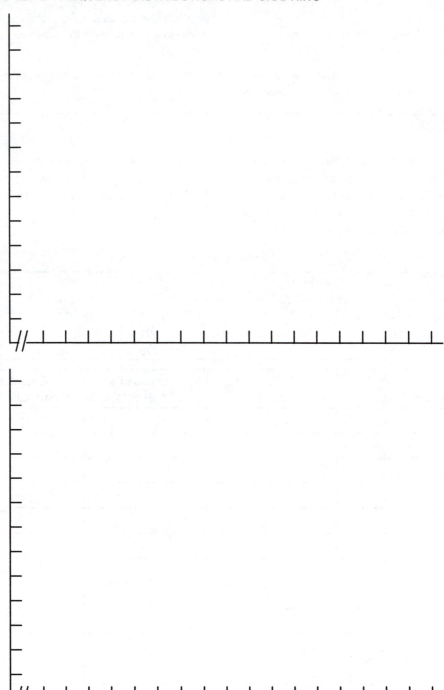

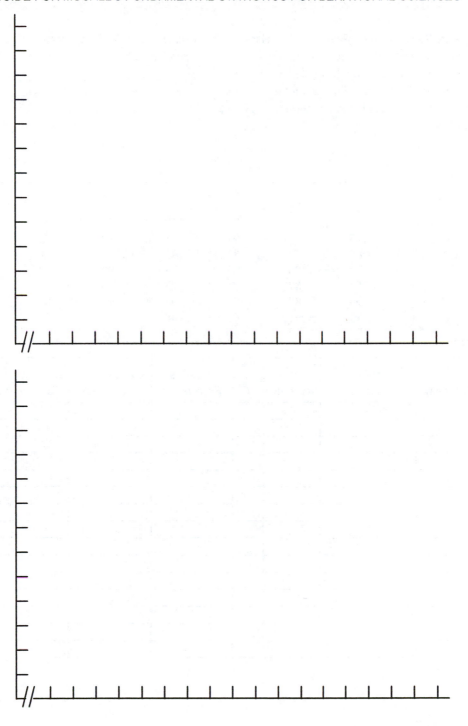

2. Below are 70 scores on a final exam in a class on Gender Roles. Follow the same directions for Exercise 1, but employ the procedures for grouped data. How would you assign the letter grades A, B, C, D, and F to these scores?

3. Create a stem-leaf display for the data in Exercise 2 above.

Scores

90	91	85	82	71	75	75
76	85	76	77	70	70	85
70	70	72	79	68	74	75
75	75	73	91	85	82	72
72	85	79	81	92	65	68
80	80	80	65	78	90	94
90	68	87	75	77	95	72
94	90	85	70	93	90	82
70	85	95	65	68	92	80
77	75	92	90	92	85	75

Class Interval	Real Limits	Midpoint	f	Relative Frequency	Cumulative Frequency	Cumulative Relative Frequency

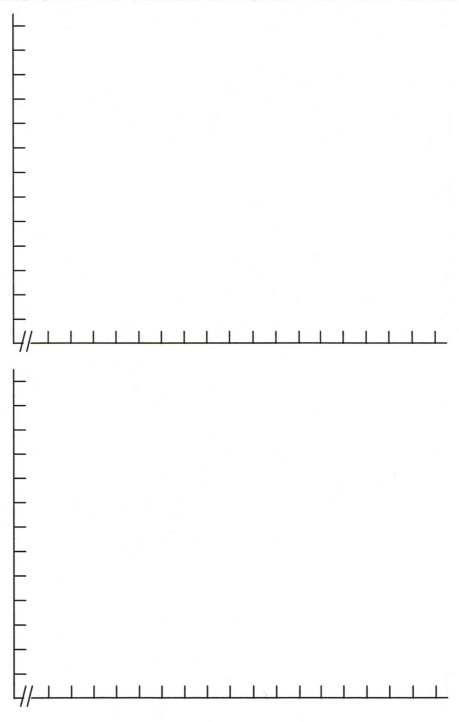

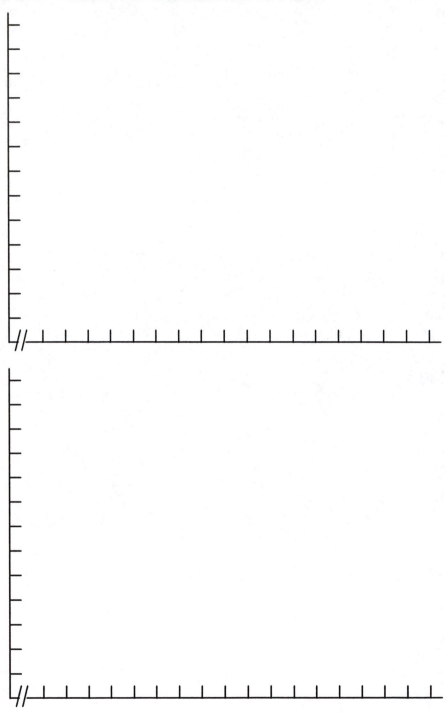

ANSWERS

Table 2-1

Score Value	Tally	f	Relative Frequency	Cumulative Frequency	Cumulative Relative Frequency
10	/	1	.05	20	1.00
9	/ / /	3	.15	19	.95
8	/ /	2	.10	16	.80
7	++++	5	.25	14	.70
6	/ / / /	4	.20	9	.45
5	/ / /	3	.15	5	.25
4	/	1	.05	2	.10
3	/	1	.05	1	.05

Table 2-4

Class Interval	Real Limits	Midpoint	f	Relative Frequency	Cumulative Frequency	Cumulative Relative Frequency
90–99	89.5–99.5	94.5	4	.08	50	1.00
80–89	79.5–89.5	84.5	8	.16	46	.92
70–79	69.5–79.5	74.5	18	.36	38	.76
60–69	59.5–69.5	64.5	9	.18	20	.40
50–59	49.5–59.5	54.5	7	.14	11	.22
40–49	39.5–49.5	44.5	4	.08	4	.08

Table 2-9

Statistics test Scores		
Depths	(Unit = 1 Point)	
($N = 50$)	Stem	Leaves
4	9	0 1 3 8
12	8	0 0 1 3 5 8 9 9
(18)	7	0 1 1 2 3 3 3 4 4 5 5 7 7 8 9 9 9 9
20	6	0 1 1 2 3 7 7 7 7
11	5	0 0 0 1 3 5 9
4	4	1 2 6 8

Self Test. (1) To summarize and describe the data. **(2)** See text pages 28-32. **(3)** See text pages 33-36. **(4a)** 89.5 and 95.5; **(4b)** 6; **(4c)** 92.5. **(5a)** T; **(5b)** F; **(5c)** T; **(5d)** F; **(5e)** T; **(5f)** F; **(5g)** T. **(6)** See text pages 42-43. **(7)** See text pages 49-55. **(8)** Variability increases. **(9)** d. **(10)** They both represent cumulative frequencies. In a cumulative frequency distribution, frequencies are cumulated only in one direction, up from the bottom or lowest interval all the way to the top or highest interval. In a stem-and-leaf display, depth is a cumulation of frequencies in both directions. One starts at the bottom and accumulates frequencies upward until reaching the line with the middle score; then one starts at the top and accumulates downward until reaching the line containing the middle score. The line containing the middle score receives, in parentheses, a depth equal to its frequency, not cumulative frequency.

Exercises .(1) See the following tables.

(1)

Temperature	f	Relative Frequency	Cumulative Frequency	Cumulative Relative Frequency
80	1	.04	25	1.00
79	1	.04	24	.96
78	1	.04	23	.92
77	4	.16	22	.88
76	5	.20	18	.72
75	4	.16	13	.52
74	3	.12	9	.36
73	3	.12	6	.24
72	0	.00	3	.12
71	2	.08	3	.12
70	1	.04	1	.04
	$N = 25$	1.00		

(2)

Class Interval	Real Limits	Midpoint	f	Relative Frequency	Cumulative Frequency	Cumulative Relative Frequency
95–99	94.5–99.5	97	2	.03	70	1.00
90–94	89.5–94.5	92	15	.21	68	.97
85–89	84.5–89.5	87	9	.13	53	.76
80–84	79.5–84.5	82	8	.11	44	.63
75–79	74.5–79.5	77	16	.23	36	.51
70–74	69.5–74.5	72	13	.19	20	.29
65–69	64.5–69.5	67	7	.10	7	.10

(3)

Final Exam Scores		
Depths ($N = 70$)	(Unit = 1 Point) Stem	Leaves
2	9	5 5
17	9*	0 0 0 0 0 0 1 1 2 2 2 2 3 4 4
26	8	5 5 5 5 5 5 5 5 7
34	8*	0 0 0 0 1 2 2 2
(16)	7	5 5 5 5 5 5 5 5 6 6 7 7 7 8 9 9
20	7*	0 0 0 0 0 0 1 2 2 2 2 3 4
7	6	5 5 5 8 8 8 8

STATISTICAL PACKAGES

StataQuest
(To accompany Exercise 1)

StataQuest will construct a simple frequency (labeled *Freq.*), relative frequency (*Percent*), and cumulative relative frequency (*Cum.*) distribution as well as draw a dotplot, a stem-and-leaf display, and a relative frequency (*Fraction*) histogram.

Start the StataQuest program. To bring up the spreadsheet for data entry, point to the *File* menu and click *New*:

> *File>New*

The Stata Editor window, which consists of a spreadsheet grid, should fill the screen.
For Exercise 1, the data are:

80 78 75 75 74 75 73 71 73 74 70 79 76 76 77 73 71 77 77 76 77 74 75 76 76

Type the scores into the first column of the spreadsheet, starting with the first value in the upper left cell. Hit the *Enter* key to register each value. Continue entering score values in the first column, until all values have been entered.

At the top of the column, double click on the default variable name, *var1*. Enter the variable name *Scores* in place of the default name, then click OK. The spreadsheet should appear as follows:

> Scores
> 80
> 78
> 75
> ⋮
> etc.

Check the data you have entered, modify if necessary, then close the Stata Editor by clicking on the *Close* button at the upper right of the Stata Editor screen. (Note that you may reopen the Stata Editor spreadsheet at any time by clicking on the *Editor* button at the top of the screen.)

You may save the data you entered as follows:

> *File>Save as*
>> (Enter drive and file name for the Stata Data save file.)
>> OK

You must open a log file to be able to save or print the results of the analysis. To open a log file, click on the *Log* button at the top left of the screen, enter a name and drive, and click OK.

Generate the frequency distribution of the scores:

> *Summaries>Tables>One-way (frequency)*
>> Data variable: Scores
>> OK

For a visual display of the distribution, you may choose the dotplot, the stem-and-leaf display, or a histogram:

> *Graphs>One variable>Dotplot*
>> Data variable: Scores
>> OK

> *Graphs>One variable>Stem-and-leaf*
>> Data variable: Scores
>> OK

> *Graphs>One variable>Histogram>Discrete variable*
>> Data variable: Scores
>> OK

The dotplot and stem-and-leaf displays are written to the log file; the histogram yields a separate graphic image.

To print the frequency tables and the first two data displays above, print the log file:

> *File>Print Log*
>> OK

To print the graph generated from the *Histogram* command:

> *File>Print Graph*
>> OK

Alternatively, the user can copy the output from the log file and paste it into a word processor. The graphic output can also be transferred to a word processor if desired. To exit StataQuest:

> *File>Exit*

StataQuest Program Output

```
. tabulate Scores

      Scores|      Freq.        Percent          Cum.
------------+-----------------------------------------
       70 |          1           4.00           4.00
       71 |          2           8.00          12.00
       73 |          3          12.00          24.00
       74 |          3          12.00          36.00
       75 |          4          16.00          52.00
       76 |          5          20.00          72.00
       77 |          4          16.00          88.00
       78 |          1           4.00          92.00
       79 |          1           4.00          96.00
       80 |          1           4.00         100.00
------------+-----------------------------------------
      Total |         25         100.00


. dplot Scores


                                  .
                       .      .   :     :
       .     :         :      :   :     :
      -+---------+---------+---------+---------+---------+-   (25 obs.)
       70        72        74        76        78        80


. stem Scores

Stem-and-leaf plot for Scores

   7* | 011
   7t | 333
   7f | 4445555
   7s | 666667777
   7. | 89
   8* | 0
```

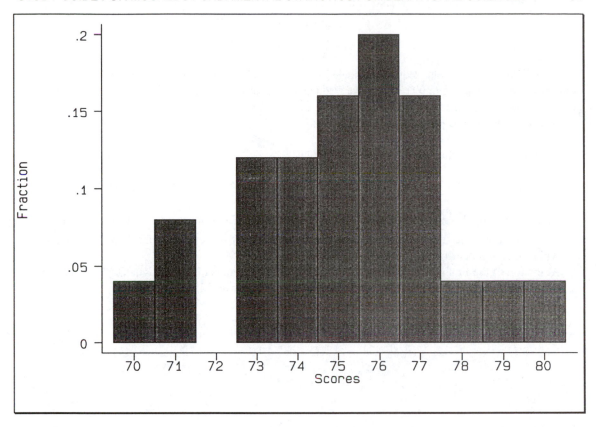

Notice that the table above is different in some respects from that presented in the text and *Guide*. First, the frequency distribution starts with the lowest score at the top, rather than the highest as in the text and *Guide*. Second, score values that have no frequencies are simply omitted, rather than listed with 0 frequencies. Third, the stem-and-leaf display has used only 6 stems, rather than all 10, abbreviated the stems as shown in the text but listed them in descending rather than ascending order similar to the frequency distribution, and omitted the depths. Also, StataQuest only produces a relative frequency histogram, which is labeled "Fraction" rather than "Relative Frequency." It does not produce a simple frequency histogram.

Minitab
(To accompany Exercise 1)

Minitab will construct simple frequency (labeled *Count*), cumulative frequency (*CumCnt*), relative frequency (*Percent*), and cumulative relative frequency (*CumPct*) distributions as well as draw dotplots, stem-and-leaf displays, and histograms.

Start the Minitab program.

For Exercise 1, the data are:

80 78 75 75 74 75 73 71 73 74 70 79 76 76 77 73 71 77 77 76 77 74 75 76 76

To enter data on the spreadsheet, click on the Data window. Enter the score values down the first column of the spreadsheet, starting with row 1. Strike the down arrow key ($\downarrow$) or the *Enter* key to register each value. Continue until all values have been entered.

At the top of the first column, click on the blank space below the column designator *C1*. Type the variable name *Scores* in this space. The spreadsheet should appear as follows:

> Scores
> 80
> 78
> 75
> ⋮
> etc.

Check the data you have entered in the spreadsheet and modify as necessary. You may save the data you have entered as follows:

> *File>Save Worksheet As*
> (Enter drive and a file name for the Minitab save file.)
> OK

Generate a frequency distribution of the scores. Using the pull-down menus, click on *Stat*, then *Tables*, then *Tally*. To choose the variable to tally, double-click on the variable *Scores* and it will be transferred across to the variables box.

> *Stat>Tables>Tally*
> Variables: Scores   [double-click to transfer]
> Display: Counts, Percents, Cumulative counts, Cumulative percents
> OK

The spreadsheet will be replaced by the Session window, on which the output will appear. The easiest way to generate a histogram is to use the descriptive statistics command. It also yields a table of statistics including the mean, standard deviation, and quartiles:

> *Stat>Basic Statistics>Descriptive Statistics*
>> Variables: Scores   [double-click to transfer]
>> Graphs:
>>> Histogram of data: Yes
>>> OK
>> OK

The histogram is a graphic output and appears in a separate graphics window. A simpler alternative to the histogram is the dotplot, which displays a dot for each score in the distribution:

> *Graph>Character Graphs>Dotplot*
>> Variables: Scores   [double-click to transfer]
>> OK

Finally, a stem-and-leaf display may be generated:

> *Graph>Character Graphs>Stem-and-leaf*
>> Variables: Scores   [double-click to transfer]
>> OK

The tables, dotplot, and stem-and-leaf outputs yield a simple character display which is written to the Session window. To print the output from the Session window:

> *File>Print Window*
>> Print Range: All
>> OK

To print the graphic image generated by the *Histogram* command:

> *Window>Hist of Scores*
> *File>Print Window*
>> Print Range: All
>> OK

Alternatively, the user can highlight the output in the Session window and copy and paste it into a word processor. The graphic output can be copied and pasted in similar fashion.

To exit Minitab:

> *File>Exit*

Minitab Program Output

Summary Statistics for Discrete Variables

Scores	Count	CumCnt	Percent	CumPct
70	1	1	4.00	4.00
71	2	3	8.00	12.00
73	3	6	12.00	24.00
74	3	9	12.00	36.00
75	4	13	16.00	52.00
76	5	18	20.00	72.00
77	4	22	16.00	88.00
78	1	23	4.00	92.00
79	1	24	4.00	96.00
80	1	25	4.00	100.00
N=	25			

Descriptive Statistics

Variable	N	Mean	Median	Tr Mean	StDev	SE Mean
Scores	25	75.120	75.000	75.130	2.438	0.488

Variable	Min	Max	Q1	Q3
Scores	70.000	80.000	73.500	77.000

Character Dotplot

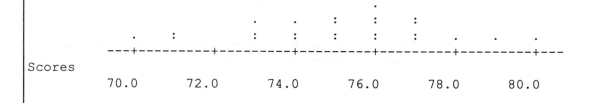

Character Stem-and-Leaf Display

```
Stem-and-leaf of Scores    N  = 25
Leaf Unit = 0.10


     1     70 0
     3     71 00
     3     72
     6     73 000
     9     74 000
    (4)    75 0000
    12     76 00000
     7     77 0000
     3     78 0
     2     79 0
     1     80 0
```

Histogram of Scores

Notice that the above output is different in some ways from that presented in the text and this *Guide*. First, scores are listed from the smallest at the top to the largest at the bottom in the frequency distribution and stem-and-leaf display. Second, score values with no frequencies are simply omitted from the frequency distribution, rather than listed with 0 frequencies. Third, each of the 11 stems are listed in the stem-and-leaf display (fewer stems could have been used), so each instance of each leaf is indicated with a 0.

SPSS
(To accompany Exercise 1)

SPSS will execute a frequency distribution, a relative frequency distribution ("Percent"), and a cumulative relative frequency distribution ("Cumulative Percent") as well as draw the histogram-like bar chart and the stem-and-leaf plot.

Start the SPSS program.

For Exercise 1, the data are:

80 78 75 75 74 75 73 71 73 74 70 79 76 76 77 73 71 77 77 76 77 74 75 76 76

The SPSS Data Editor, which consists of a spreadsheet grid, should fill the screen. You must enter the data values into the first column of the spreadsheet. Type the first value in the upper left cell, hit *Enter* or the down arrow key, then wait briefly until SPSS sets up the variable. The first value will appear in the cell.

Continue entering the scores down the first column until all values have been entered. At the top of the column, double click on the default variable name, *var0001*. Enter the new variable name *Scores* in place of the default name, and click OK. The spreadsheet should appear as follows:

Scores
80
78
75
⋮
etc.

Check the data you entered, modify if necessary, then save the data as follows:

File>Save as
(Enter device and file name for SPSS save file.)
Save

Generate the frequency distribution of the scores along with a histogram or bar chart:

Statistics>Summarize>Frequencies
Variable(s): Scores
(Click on the arrow to transfer Scores to the variables box.)
Charts:
Chart type: Bar chart(s)
Continue
OK

Note that the chart type "Bar chart" will produce a bar for each distinct value of Scores, and only for intervals with one or more counts. The interval around 72, for instance, is not presented on the bar chart because the count is zero for this score.

To generate the stem-and-leaf display of Scores:

> *Statistics>Summarize>Explore>*
> > Dependent List: Scores
> > Display: Plots
> > Plots:
> > > Boxplots: None
> > > Descriptive: Stem-and-leaf
> > > Continue
> > OK

Print the frequency tables, the bar chart, and the stem-and-leaf display:

> *File>Print*
> > Print range: All visible output
> > OK

Exit SPSS:

> *File>Exit SPSS*

SPSS Program Output

Frequencies

Statistics

	N	
	Valid	Missing
SCORES	25	0

Scores

		Frequency	Percent	Valid Percent	Cumulative Percent
Valid	70.00	1	4.0	4.0	4.0
	71.00	2	8.0	8.0	12.0
	73.00	3	12.0	12.0	24.0
	74.00	3	12.0	12.0	36.0
	75.00	4	16.0	16.0	52.0
	76.00	5	20.0	20.0	72.0
	77.00	4	16.0	16.0	88.0
	78.00	1	4.0	4.0	92.0
	79.00	1	4.0	4.0	96.0
	80.00	1	4.0	4.0	100.0
	Total	25	100.0	100.0	
Total		25	100.0		

Bar Chart

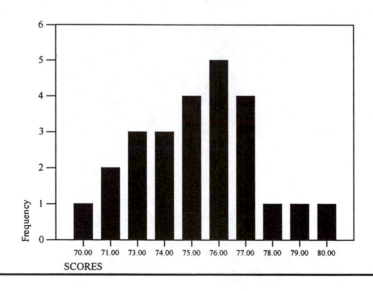

Explore

Case Processing Summary

	Cases					
	Valid		Missing		Total	
	N	Percent	N	Percent	N	Percent
SCORES	25	100.0%	0	.0%	25	100.0%

Scores

SCORES Stem-and-Leaf Plot

Frequency   Stem & Leaf

```
  3.00     7 . 011
  3.00     7 . 333
  7.00     7 . 4445555
  9.00     7 . 666667777
  2.00     7 . 89
  1.00     8 . 0
```

Stem width: 10.00
Each leaf: 1 case(s)

Notice that the above output is different in some ways from that presented in the text and this *Guide*. First, scores are listed from the smallest at the top to the largest at the bottom in the frequency distribution and stem-and-leaf display. Second, score values with no frequencies are simply omitted from these two presentations, rather than listed with 0 frequencies. Third, the stems actually represent two leaves each, but the designations of *, t, f, s, used to represent that fact are omitted, and "Frequency" is used instead of the word "Depth."

CHAPTER 3

CHARACTERISTICS
OF DISTRIBUTIONS

CONCEPT GOALS

Be sure that you thoroughly understand the following concepts and how to use them in statistical applications.

- ◆ Mean; $\sum (X_i - \overline{X}) = 0$ and $\sum (X_i - \overline{X})^2$ is a minimum

- ◆ Median

- ◆ Mode; bimodal distribution

- ◆ Advantages of mean, median, and mode

- ◆ Range

- ◆ Variance and standard deviation

- ◆ Resistant indicators; fourth-spread, outliers, extreme scores; five-number summary, boxplot.

- ◆ Population parameters and sample statistics

GUIDE TO MAJOR CONCEPTS

Central tendency and **variability** are major concepts in statistics. There are several numerical ways to express the concept of the typical score in a distribution, more formally known as the _____ _____ of a distribution. Other numerical indices express the extent to which scores in a distribution differ from one another and from their central tendency, that is, they reflect the amount of _____ in a distribution.

central tendency

variability

Central Tendency

There are at least three types of typical scores, or indicators or measures of _____ _____ , which are the **mean**, **median**, and **mode**. The most common is the arithmetic average, or _____ , which is defined as the sum of the scores divided by the number of scores in the distribution. The mean is symbolized by _____ , and its formula can be expressed symbolically as _____ . If a distribution consists of the scores (2, 3, 6, 9, 5), the mean is given by

$$\overline{X} = \underline{\quad\quad} = \underline{\quad\quad} = \underline{\quad\quad} .$$

central tendency

mean

$\overline{X}$

$\dfrac{\sum X_i}{N}$

$\dfrac{\sum X_i}{N} = \dfrac{25}{5} = 5$

The mean has two characteristics that make it the most commonly used measure of central tendency. The first is that in a distribution the sum of the deviations of the scores about their mean always equals _____ . If X_i is any particular score in the sample and $\overline{X}$ is the mean of all the X_i, this relationship can be symbolically written _____ . Two distributions are given in Table 3-1 with space for you to calculate the mean as well as the sum of the deviations of scores about their mean, $\sum (X_i - \overline{X})$. Perform these calculations for the X and the Y distributions. Now select any value k that is not equal to the mean of the Y distribution and calculate the deviations of the Y_i scores about the value k instead of about the mean. What do you notice about the sum of the deviations when they are taken about the mean rather than about some other value? The sum about the mean is _____ .

0

$\sum (X_i - \overline{X}) = 0$

0

Now calculate the columns of Table 3-1 that require the *squared* deviations of scores about their mean and about k. For the X_i, notice that while the sum of deviations about the mean of a distribution is always ___ , when the deviations are squared their sum is ___ ,

0; 38

that is, not 0. These two principles can be written in symbols as

$$\underline{\hspace{2cm}} = \underline{\hspace{1cm}}$$

$$\underline{\hspace{2cm}} \neq \underline{\hspace{1cm}}$$

$$\sum (X_i - \overline{X}) = 0$$

$$\sum (X_i - \overline{X})^2 \neq 0$$

The second point to observe about the mean is that the sum of *squared* deviations about the mean, while not usually _____ , is _____ than the sum of squared deviations about any other value, such as the k you selected. The sum of squared deviations about the mean for distribution Y is ___ , but the sum of squared deviations about k is _____ , illustrating the fact that the sum of the _____ deviations about the mean is _____ than the sum of squared deviations about any other value. It is in this sense, sometimes called the **least squares sense**, that the mean is closer to the scores (in terms of squared deviations) than is any other measure of central tendency (any other k.)

0

smaller

70

(depends upon your k, but > 70)

squared; smaller

To review, the mean is defined by the formula _____ , and it possesses two characteristics that make it a good measure of central tendency. These may be stated symbolically as follows:

$$\underline{\hspace{3cm}} = \underline{\hspace{3cm}} \text{ and } \underline{\hspace{2cm}} \text{ is a minimum.}$$

$$\overline{X} = \sum X_i / N$$

$$\sum (X_i - \overline{X}) = 0; \sum (X_i - \overline{X})^2$$

Although the mean is the most common and the best (in the least squares sense) index of central tendency, it is not the only index, nor is it always the most appropriate one to use. For example, the point that divides the distribution into two parts such that an equal number of cases lie above and below that point is the _____ , symbolized by _____ . The calculation of the median varies, depending on whether there is an odd or an even number of total cases (i.e., N) in the distribution.

median

M_d

1. **If there is an odd number of cases in the distribution (i.e., N is odd),** the median is the score value corresponding to the middle case, which will be the $(N+1)/2$ case from the bottom of the distribution. For example, if the distribution is

$$(5, 7, 8, 9, 11)$$

there is an odd number of cases, namely $N =$___ , so the middle case will be $(N+1)/2 =$_____ $=$_____ or the third case from the bottom. The score value of the third case is _____ , so the _____ is 8. Note that this score value divides the distribution into two equal parts, since two cases fall below and two cases fall above the score value of 8.

5

$(5+1)/2 = 3$

8; median

Table 3-1 Means and Deviations

Distribution X				Distribution Y						
	Deviations about $\bar{X}$				Deviations about $\bar{Y}$				Deviations about k	
X_i	$\bar{X}$	$(X_i - \bar{X})$	$(X_i - \bar{X})^2$	Y_i	$\bar{Y}$	$(Y_i - \bar{Y})$	$(Y_i - \bar{Y})^2$	k	$(Y_i - k)$	$(Y_i - k)^2$
2				1						
7				2						
2				8						
9				9						
5				10						
$\sum X_i =$		$\sum(X_i - \bar{X}) =$	$\sum(X_i - \bar{X})^2 =$	$\sum Y_i =$		$\sum(Y_i - \bar{Y}) =$	$\sum(Y_i - \bar{Y})^2 =$		$\sum(Y_i - k) =$	$\sum(Y_i - k)^2 =$

$N=$ _____

$\bar{X}=$ _____

$N=$ _____

$\bar{Y} =$ _____

If the distribution is

$$(19, 21, 27, 32, 32, 41, 55)$$

again there is an ____ number of cases (i.e., $N = 7$), the middle case   odd
is the _____ = _____ = _____ case, and the value of the $(N+1)/2 = (7+1)/2 = 4\text{th}$
median is _____ . This shows that the median is the value of the   32
middle case even when more than one case has that score value.

2. If there is an even number of cases in the distribution, the
median is half way between (i.e., is the average of) the score values
of the two middle cases, which will be cases $N/2$ and $(N/2)+1$ from

the bottom of the distribution. Suppose the distribution is

$$(4, 6, 8, 10, 12, 15)$$

Here there is an even number of cases ($N = $ ___) and the two middle   6
cases are the $N/2 =$ _____ case and the $(N/2)+1 =$ _____ $6/2 = 3\text{rd}; (6/2)+1 = 4\text{th}$
case. The median is the average of the score values of these two
middle cases, or _____ = _____ . Notice that the point 9, while $(8+10)/2 = 9$
not actually in the distribution, would separate the distribution into
_____ _____ parts consisting of three cases below and three   two equal
cases above this point.

Now consider the distribution

$$(12, 14, 16, 19, 19, 22, 27, 30)$$

Here, $N = 8$ is _____ , and the middle cases are the   even
_____ = _____ = _____ and _____ = $N/2 = 8/2 = 4\text{th}; (N/2)+1 =$
_____ = __ . The median is the average of their score values, which is $8/2+1 = 5\text{th}$
simply ___ . This shows that the median is still the average of the   19
values of the middle cases even when those cases have the same
score value.

Table 3-2 summarizes these steps in determining the median at
the left and then provides three additional examples. Determine the
medians for these three distributions now.

The third index of central tendency is the easiest one to compute.
It is the most frequently occurring score, called the _____ and   mode
symbolized by _____ . In the distribution (2, 4, 6, 6, 6, 5, 5, 10), the M_o
most frequent score is _____ , which is called the _____ .   6; mode

We have examined three indices of central tendency: One is the
score value closest in the least squares sense to the scores in the
distribution (the _____); one divides the distributions into two   mean
equally-sized parts (the _____); and one is the most frequent   median
score (the _____). The values of these three indices are not   mode
usually the same for a particular distribution. For example, consider
the four distributions in Table 3-3. Calculate the mean, median, and
mode for each, and then locate and mark each index of central
tendency with its symbol on the scale of measurement corresponding
to each distribution in the bottom half of the table.

Table 3-2 Computational Examples for Determining the Median (M_d)

	Distribution A	Distribution B	Distribution C
0. Raw distribution:	(4, 10, 2, 15, 1, 3, 11, 9, 4, 3, 10, 14, 11)	(4, 2, 6, 8, 2, 6, 5, 3, 1, 2, 1, 8, 0, 7)	(2.1, 1.9, 2.8, 2.4, 1.6, 2.5, 2.1, 2.4, 1.9, 2.1, 2.0)
1. Order all the cases by score value:	( )	( )	( )
2. Determine N:	$N=$ ___	$N=$ ___	$N=$ ___
3. Determine location of M_d:	M_d is at/between case(s)	M_d is at/between case(s)	M_d is at/between case(s)
a. If N odd, M_d is the value of case $(N+1)/2$	___	___	___
b. If N even, M_d is the average of cases $N/2$ and $(N/2)+1$	___	___	___
4. Determine the value of M_d:	$M_d =$ ___	$M_d =$ ___	$M_d =$ ___

Table 3-3 Comparison of the Mean, Median, and Mode

	W	X	Y	Z
	2	2	3	3
	3	3	3	5
	5	5	5	5
	6	6	7	7
	9	24	7	

Mean

Median

Mode

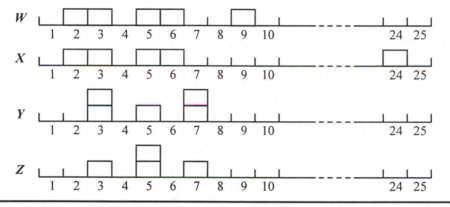

Look first at distributions *W* and *X*. Since there is no duplication of scores in either distribution, there is no useful _____ . Observe that the two distributions are identical except for the last score in each, 9 in *W* and 24 in *X*. The difference between these two cases is reflected in a change in the _____ , but it has no effect on the value of the _____ . The formula for the mean, _____ , indicates that the value of *each* score is entered and has an effect on this statistic. Thus, the measure of central tendency that acts most like a fulcrum or balance point for the distribution of score values along the measurement scales at the bottom of Table 3-3 is the _____ . In contrast, the point that balances cases, not score values, by dividing the cases into two sets having an equal number of cases is the _____ . Only the score value of the *central* cases, regardless of how far the other score values are from them, influences the _____ . Therefore, in distributions that include one or two extreme score values which you regard as atypical, you might prefer the _____ over the _____ as a descriptive index of _____ _____ . Another way of saying the same thing is to observe that distribution

mode

mean

median; $\sum X_i / N$

mean

median
median

median
mean; central tendency

X can be described as having positive **skewness**; when distributions
are asymmetrical or seriously _____ , one may prefer the   skewed
_____ over the _____ , because the skewness will   median; mean
have less influence on the _____ than on the _____ .   median; mean

In determining the mode of distribution Y, you will have noticed
that this distribution is _____ . When scores tend to bunch in   bimodal
two places, the mean and median do not reflect this fact but the
_____ does.   mode

Distribution Z is a simple example of a symmetrical distribution
with one mode. In this case the mean, median, and mode are _____ .   equal

Variability

Distributions do not differ only with respect to central tendency;
they also differ with respect to **variability**. For example, the two
distributions (3, 4, 5) and (1, 4, 7) have the same mean and median,
but they differ in the extent to which the scores deviate from one
another and from their central tendency, a characteristic called
_____ .   variability

One measure of variability is simply the difference between the
largest and smallest score, a statistic known as the _____ .   range
Thus, for the two distributions given above, the ranges are ____ and   2
____ . Although the range provides some idea of the _____ of   6; variability
the scores, it ignores cases less extreme in value than the smallest and
largest values. The imprecision of the range as a measure of
variability is illustrated by the fact that for (1, 4, 7, 11, 14) and (1, 7,
7, 7, 14) the two ranges are (equal/unequal) _____ , but the first   equal
distribution has (more/less) _____ variability than the second   more
because the score values in it deviate more from each other and from
their central tendency.

The most common numerical index of variability is the _____   variance
symbolized by ____ and defined by the formula _____ . s^2; $\sum(X_i - X)^2/(N-1)$
The square root of the variance, symbolized by _____ and called the s
_____ _____ , is also used as an index of   standard deviation
variability. The formula for the variance states that the variance
equals the sum of _____ _____ of the scores   squared deviations
about their _____ divided by the quantity _____ .   mean; $N-1$
Thus the variance is roughly akin to the average squared deviation
between scores and their mean.

To understand that the variance indeed reflects the degree to
which scores deviate from one another, consider Table 3-4. For each
of the three distributions, first calculate the mean, the deviations

about the mean $(X_i - \overline{X})$, and the squared deviations about the mean $(X_i - \overline{X})^2$; then calculate each variance. Now, to see that the variance is proportional to the sum of the squared deviations between each score and every other score, the second part of the table helps you calculate these values. The pairs of scores within each distribution are already listed for you; square the differences between the numbers in each pair, and add them up to give you the sum of squared differences between each score and every other score, symbolized in the table by $\sum d^2$. It happens that the variance equals that sum divided by twice the number of such pairs of scores, which we have symbolized by $2n_d$ and which is 6 for these particular distributions. If you have not made an error, s^2 and $\sum d^2/2n_d$ will be (equal/unequal) _____ . Write the value of s^2 for each distribution in the space provided at the right of the measurement scales drawn at the bottom of the table, and then determine $\sqrt{s^2} = s$ and write it in the available space. Notice that as the scores in the distribution become more spread out (that is, as they deviate more from one another), the variance (increases/decreases) _____ . This illustrates how s^2 (and its cousin, s) reflect the _____ in a distribution.

      equal

      increases

      variability

    Table 3-5 presents a guided computational example for the concepts presented above. Complete it now and use it as a model for future computations.

Resistant Indicators

The indices of central tendency and variability presented above are the most traditional and common characteristics used to describe distributions, but they have limitations. One limitation is that the value of many of these indicators depends on the value of every score in the distribution. Ordinarily, this feature is valuable, but if one or two scores are unusually high or low, the mean, range, variance, and standard deviation also will be unusually high or low. It would be helpful to have **resistant indicators**, ones that change relatively little in value if a small portion of the data is replaced with new numbers that may be very different than the original ones. Presumably, _____ _____ "resist" influence by individual scores, especially atypical and very deviant scores, and therefore reflect only the main body of the data.

      resistant indicators

Table 3-4 Variability and the Variance

Distribution X				Distribution Y				Distribution W			
X_i	$\overline{X}$	$(X_i - \overline{X})$	$(X_i - \overline{X})^2$	Y_i	$\overline{Y}$	$(Y_i - \overline{Y})$	$(Y_i - \overline{Y})^2$	W_i	$\overline{W}$	$(W_i - \overline{W})$	$(W_i - \overline{W})^2$
5				3				1			
6				6				6			
7				9				11			

$\sum X_i =$ $\sum (X_i - \overline{X})^2 =$
$N =$
$\overline{X} =$ $N - 1 =$
     $s_x^2 =$

$\sum Y_i =$ $\sum (Y_i - \overline{Y})^2 =$
$N =$
$\overline{Y} =$ $N - 1 =$
     $s_y^2 =$

$\sum W_i =$ $\sum (W_i - \overline{W})^2 =$
$N =$
$\overline{W} =$ $N - 1 =$
     $s_w^2 =$

Average Squared Differences between Each Pair of Scores

$(7-6)^2 =$

$(7-5)^2 =$

$(6-5)^2 =$

$\sum d_x^2 =$

$2n_d = 6$

$\sum d_x^2 / 6 = \stackrel{?}{=} s_x^2$

$(9-6)^2 =$

$(9-3)^2 =$

$(6-3)^2 =$

$\sum d_y^2 =$

$2n_d = 6$

$\sum d_y^2 / 6 = \stackrel{?}{=} s_y^2$

$(11-6)^2 =$

$(11-1)^2 =$

$(6-1)^2 =$

$\sum d_w^2 =$

$2n_d = 6$

$\sum d_w^2 / 6 = \stackrel{?}{=} s_w^2$

s^2 _____ _____ _____

X: 1 2 3 4 5 6 7 8 9 10 11

Y: 1 2 3 4 5 6 7 8 9 10 11

W: 1 2 3 4 5 6 7 8 9 10 11

mean

Table 3-5 Guided Computational Example

X_i	X_i^2		Central Tendency
5		1.	$\bar{X} = \dfrac{\sum X_i}{N} =$
7		2.	M_d = value of middle score =
4			$M_d =$ =
5		3.	M_o = value of most frequent score
9			$M_o =$
6			

$\sum X_i =$ $\sum X_i^2 =$

$(\sum X_i)^2 =$

$N =$

Variability

1. Range = largest minus smallest score value =

2. $s^2 = \dfrac{\sum (X_i - \bar{X})^2}{N-1} = \dfrac{N \sum X_i^2 - (\sum X_i)^2}{N(N-1)} =$ =

$s = \sqrt{s^2} =$ =

One _____ _____ of central tendency is
the median. Recall from Table 3-3 that the value of every score in the
distribution, including very deviant scores, influences the value of the
_____ , but deviant scores do not influence the value of the
_____ . In the following two distributions

A: (2, 4, 6)
B: (2, 4, 100)

The mean of A is ____ , while the mean of B is _____ ; but the
median of both A and B is ____ . So the deviant score of 100
influences the value of the _____ but not the _____ ;
consequently, the median is a more _____ _____ of
central tendency.

The traditional indicators of variability, namely the _____ ,
_____ , and _____ _____ , are also
influenced by the value of every score and especially the value of any

resistant indicator

mean
median

4; 35.3
4
mean; median
resistant indicator

range
variance; standard deviation

very deviant scores. In the above distributions, for example, the range of A is ____ – _____ = _____ and the range of B is ____ – ___ = _____ . The variance and standard deviation would also be influenced by this very deviant score. Therefore, it would be helpful to have an indicator of variability that was more _____ to the value of unusual and deviant scores.

$6-2=4$
$100-2=98$

resistant

One such indicator is called the **fourth-spread**, and its definition is based upon the concept of **depth** in **stem-and-leaf displays** presented in the previous chapter. Table 3-6 presents a stem-and-leaf display for the number of looks at mother that $N = 70$ ten-month infants made during a 20-minute free play session in a study of attachment. At the left of the table are the depths. Recall that the _____ of a line in a _____-_____-_____ _____ is its cumulative frequency from the bottom or top of the display, whichever is smaller. So in Table 3-6, the depth of line 1* is _____ , which means that 26 of the 70 cases have scores of ____ or _____ . Similarly, for the line 3*, the _____ is 9, which means that _____ of the ____ scores in the batch are ____ or _____ . Recall also that the depth of (13) for line 1· means that this line contains ___ cases, and that it contains the middle score value, or _____ .

depth; stem-and-leaf display

26; 14
lower; depth
9; 70; 30; higher

13
median

The median score value has a depth of $(N+1)/2$. If N is an odd number, this depth will be a whole number, and the median is the value of the case having that depth. That is, if the depth of the median is 11, the median is the value of the ____ score from the bottom (or top.) If N is an even number, the depth of the median will be fractional. For example, the batch in Table 3-6 has an $N = 70$, so the depth of the median is $(N+1)/2 =$ _____ = _____ . The median is the average of the scores bordering this fractional depth; that is, the median is the average of the values of the ____ and ____ scores, or (_____ + _____)/_____ = _____ .

11th

$(70+1)/2 = 35.5$

35th; 36th
$(18+18)/2 = 18$

Just as there is a depth of the median, there are also **depths of the fourths**. Whereas the median is the middle score value that divides the distribution or **batch** in half, the fourths (plus the median) divide the _____ approximately into _____ . The depth of the fourths is given by

$$\text{depths of the fourths} = \frac{(\text{depth of median}) + 1}{2}$$

batch; fourths

in which a fractional depth of the median is first rounded down to the next lowest integer. In the present example, the depth of the median was ____ , which must be rounded down to ____ before determining the depth of the _____ , which is (___+___)/___=___ . Therefore, in this batch, the **fourths** are the values of the _____ score

35.5; 35
fourths; $(35+1)/2 = 18$
18th

Table 3-6 Guided Computational Example of Resistant Indicators

	Number of Looks at Mother	
Depths	(Unit = 1 Look)	
($N = 70$)	Stem	Leaves
1	4·	7
1	4*	
2	3·	5 5
9	3*	0 1 1 2 4 4
18	2·	5 7 7 7 8 8 8 8 9
31	2*	0 0 1 1 1 2 2 2 2 3 4 4 4
(13)	1·	5 5 5 6 6 6 7 8 8 8 9 9
26	1*	0 0 1 1 1 2 2 2 2 3 4 4
14	0·	5 6 6 6 6 7 8 8 9 9
4	0*	0 4 4 4

Median

The **median** is the value of the score having a depth of $(N+1)/2$. The median has a depth of $(N+1)/2 =$ _____ = _____.

   If the depth of the median is a whole number, the value of the median is the value of the case having this depth. If the depth of the median is fractional, the value of the median is the average of the values of the cases bordering this fractional depth.

$$M_d = \underline{\hspace{3cm}}$$

Fourth-Spread

Depth of the fourths $= \dfrac{(\text{depth of the median}) + 1}{2}$ in which any fractional depth of the median is first rounded down to the next whole number.

$$\text{Depth of the fourths} = \frac{(\quad\quad) + 1}{2} = \underline{\hspace{2cm}}.$$

   The **lower fourth**, F_L, is the score value of the case with a depth of a fourth counted up from the bottom of the batch.

F_L is the value of _____ case from the bottom which is $F_L =$ _____ .

   The **upper fourth**, F_U, is the score value of the case with a depth of a fourth counted down from the top of the batch.

F_U is the value of _____ case from the top which is $F_U =$ _____ .

   The **fourth-spread** is the difference in values between the upper and lower fourths:

fourth-spread = upper fourth − lower fourth
fourth-spread = $F_U - F_L$
fourth-spread = _____ = _____

Outliers and Extremes

Outliers are cases whose values fall either
below $F_L - 1.5$ (fourth-spread) =_____ =_____
or
above $F_U + 1.5$ (fourth-spread) =_____ =_____
So the score value of _____ is an outlier.

Extreme scores are the lowest and highest scores in the batch excluding outliers.
The extreme scores are LEx _____ and UEx _____ .

---

from the top, called the **upper fourth**, F_U , and from the bottom,
called the **lower fourth**, F_L . The value of the 18th case from the
bottom is in the line with a stem of ____ . Since 14 cases exist below   1*
this line, one needs four more cases from line 1*. Counting cases
(i.e., leaves) from left to right (i.e., in increasing value) along line 1*,
the fourth case has a leaf of _____ representing a score value of   1
_____ , so the value of the _____ _____,   11; lower fourth
symbolized by _____ , is ____ .Now, to obtain the 18th case from the F_L ; 11
top, we note that the depth of line 2· is 18, which means that this line
contains the 18th case and it is the last case. But remember that you
are counting *down* over *decreasing* score values, so you must count
down from right to left within a line. Thus, the 18th case has a leaf of
____ representing a score value of ____ , so the value of the _____   5; 25; upper
_____ , or __ , is __ . If the depth of a fourth is fractional (e.g., 9.5),   fourth; F_U ; 25
the fourth is the average of the scores bordering the fraction (e.g., the
average of the values of the __ and __ cases.)   9th; 10th

Now the resistant indicator of variability is the **fourth-spread**,
which is the difference in score values between the upper and lower
fourths:

$$\text{fourth-spread} = \text{upper fourth} - \text{lower fourth}$$
$$\text{fourth - spread} = F_U - F_L$$

In the present example, the _____-_____   fourth-spread
equals ____-_____=_____ .   25 − 11 = 14

The fourth-spread approximately defines the middle half of the
batch. Thus, roughly half the infants in this example looked at their
mothers between __ and ____ times. Because it is the middle half of   11; 25
the cases and it is based upon the median not the mean, the _____-   fourth-
_____ is not influenced by unusual deviant scores and is therefore   spread
a _____ _____ of variability.   resistant indicator

The fourths help define **outliers**, cases whose values are
substantially deviant from the group and would have little likelihood
of being obtained again if a new sample were taken. Outliers are
cases that fall 1.5 times the fourth-spread above or below the fourths,
that is,

below $F_L - 1.5$ (fourth-spread) or

above $F_U + 1.5$ (fourth-spread)

Recall that the fourth-spread for this example was ___ , $F_L =$ ___ ,   14; 11
and $F_U =$ ___ . So the cutoffs for _____ are: 25; outliers

$F_L - 1.5$ (fourth-spread) = _____ = _____   11 − 1.5(14) = −10

and

$F_U + 1.5$ (fourth-spread) = _____ = _____   25 + 1.5(14) = 46

Since there can be no negative number of looks at mother, the lower
fourth is taken to be, not −10, but ___ . Note that these cutoffs of   0
___ and ___ define one score as an _____ , namely the score   0; 46; outlier
value of ___ .   47

Extreme scores are the lowest and highest scores in the batch
excluding outliers. Eliminating the outlier score of 47, the next
highest score was ___ and the lowest score was ___ , which are the   35; 0
_____ _____ of the batch. They are   extreme scores
represented by *LEx* for the lower extreme score and *UEx* for the
upper extreme score. Thus, $LEx =$ ___ and $UEx =$ ___ .   0; 35

The lower portion of Table 3-6 provides a guided computational
example of these resistant indicators. Complete it now.

These indicators can be presented in a special table or
graphically. A table containing the major resistant indicators of
central tendency and variability is called a **five-number summary**.
Table 3-7 presents the general format for the five-number summary
for the information obtained in Table 3-6. Study the notation under
the display and then fill in the values from Table 3-6 in their proper
places in the _____-_____ _____ at the top of   five-number summary
Table 3-7.

A graphical display of the same information is called a **boxplot**,
and a guide for constructing a boxplot is given at the bottom of
Table 3-7. Follow those instructions and label the sample _____   boxplot
at the bottom of the table.

Notice the nice picture the boxplot provides of the nature of the
scores in the batch. The box, which spans the _____ , represents   fourths
the central _____ of the batch. It is the main body of data exclusive   half
of the upper and lower quarters of the batch that can contain deviant
scores. The resistant indicator of central tendency, the _____ ,   median
resides within the box. The ends of the lines are defined by the
_____ and _____ _____ scores. This is basically the range of   lower; upper extreme
the batch ignoring _____ , which are presented for completeness.   outliers
Asymmetry in the plot, conveyed by a longer line to one versus the
other side of the box, would imply skewness. In this case, the batch
appears symmetrical except for the _____ .   outlier

Table 3-7 Guided Examples of Presentations of Resistant Indicators

Five Number Summary

$N =$ _____

$M =$ _____

$F =$ _____

1

Measure = _____

$M_d =$ _____

$F_L =$ _____ $F_U =$ _____

$LEx =$ _____ $UEx =$ _____

in which $N =$ The total number of cases in the batch

Measure =   The name of the measurement

$M =$ The depth of the median

$M_d =$ The value of the median

$F =$ The depth of the fourths

$F_L =$ The value of the lower fourth

$F_U =$ The value of the upper fourth

$l =$ The depth of the extreme scores ignoring outliers

$LEx =$ The value of the lower extreme score

$UEx =$ The value of the upper extreme score

Boxplot

1. Mark off abscissa in units of the measurement scale and label. Draw an ordinate without scale and label the batch(es) to the left of the ordinate.
2. Draw a rectangular box for each batch that stretches between the lower fourth (F_L) and upper fourth (F_U).
3. Within the box, draw a vertical line at the median (M_d).
4. Draw horizontal lines in each direction from the box, ending at the lower extreme score (*LEx*) and at the upper extreme score (*UEx*).
5. Locate each outlier, if any, with an *x* and label that case, if appropriate (e.g., with initials).

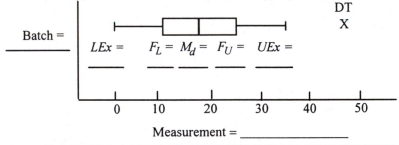

Statistics and Parameters

Frequently, research is conducted on a small group of subjects, called a _____ , but the results are generalized to a much larger group, called the _____ . A subset of a larger _____ is a _____ . Quantities calculated on a sample are called _____ and symbolized with Arabic letters, such as $\overline{X}$ for the mean and s^2 for the variance of a sample. Quantitative characteristics of populations are called _____ and symbolized with Greek letters such as μ (read "mew") for the mean and σ^2 (read "sigma squared") for the variance of a population. Often sample _____ are used to estimate the values of population _____ .

sample

population; population

sample; statistics

parameters

statistics
parameters

SELF-TEST

1. In what sense is the mean closest to the scores of a distribution?

2. What advantages does the median have over the mean as an index of central tendency?

3. Which index of central tendency assumes an equal-interval scale?

4. For each of the following distributions, indicate which measure of central tendency might be preferred and explain why.
   a. age of students in an urban university
   b. marital status (e.g., single, married, not married but living together as a couple, divorced) of students in a particular college
   c. height of all male students in a particular college

5. Which of the indices of central tendency is not influenced by a single extreme score?

6. Why is the range an imprecise measure of variability?

*7. Why is the standard deviation often preferred to the variance as a descriptive measure of variability?

8. If all the scores in a distribution are negative, will the variance be positive or negative? Explain.

9. What will be the variance if all the scores in a distribution are the same? What will be the standard deviation?

*10. Define *population* and *sample*.

*11. Why does one divide by $N - 1$ rather than by N when calculating the variance?

*12. What are the advantages of using stem-and-leaf displays, five-number summaries, and boxplots over more traditional descriptive methods?

Questions preceded by an asterisk can be answered on the basis of the discussion in the text, but the discussion in this Study Guide does not answer them.

EXERCISES

1. Calculate the mean, median, mode, range, variance, and standard deviation according to the format of Table 3-5 for each of the following distributions.

X	Y	Z
3	4	3
4	1	1
6	4	9
7	5	2
8	6	7
5	7	5
9	8	5
		5
		9
		4

2. Determine the median following the steps in Table 3-2 for the following distributions.
   a. (4, 2, 24, 1, 3)
   b. (2, 5, 8, 1, 3, 6)
   c. (2, 5, 8, 5, 5, 7)
   d. (3, 5, 5, 6, 8)
   e. (23, 25, 25, 25, 27, 37)
   f. (2.4, 2.1, 7.0, 6.2, 4.6, 3.7, 4.6)

3. Draw distributions such that the following are true.
   a. The median is greater than the mean.

   b. The mode is greater than the mean and the mean is less than the median.
   c. The mean, median, and mode are identical.

4. Calculate the variance with the computational formula given in Table 3-5 for each of the following distributions, and discuss the differences in the s^2 obtained relative to the difference in the variability of scores between the four distributions.
   a. (7, 6, 5, 7, 8, 3)
   b. (−3, −4, −5, −3, −2, −7)
   c. (17, 16, 15, 17, 18, 13)
   d. (70, 60, 50, 70, 80, 30)

5. Two basketball teams will meet for the league championship. They have played essentially the same teams during the season and they have identical records going into the championship game. Sports writers are arguing that Allegheny is a better team than Laurel Highlands because they have scored a higher average number of points. Below are stem-and-leaf displays of the points scored by the two teams. Determine five-number summaries for each and construct a single boxplot comparing the two teams. Interpret this information in view of the sports writers' claim.

	Allegheny Points per Game			Laurel Highlands Points per Game	
Depths (N = 30)	(Unit 1 = Points) Stem	Leaves	Depths (N = 30)	(Unit 1 = Point) Stem	Leaves
3	9·	7 9 9	6	8·	5 6 6 7 7 8
3	9*		13	8*	0 2 3 3 3 3 4
6	8·	6 7 8	(6)	7·	5 6 6 7 7 9
12	8*	0 1 2 2 3 4	11	7*	0 2 2 2 3 4 4
(7)	7·	5 5 7 8 8 9 9	4	6·	5 7
11	7*	0 2 3 3 4 4	2	6*	
5	6·	5 6 8 8	2	5·	9
1	6*	4	1	5*	1

ANSWERS

Table 3-1. $\sum X_i = 25$, $\overline{X} = 5$, $\sum (X_i - \overline{X}) = 0$, $\sum (X_i - \overline{X})^2 = 38$; $\sum Y_i = 30$, $\overline{Y} = 6$, $\sum (Y_i - \overline{Y}) = 0$, $\sum (Y_i - \overline{Y})^2 = 70$; $\sum (Y_i - k)^2 > \sum (Y_i - \overline{Y})^2$.

Table 3-2. Distribution A: (1, 2, 3, 3, 4, 4, 9, 10, 10, 11, 11, 14, 15), $N = 13$, M_d at case 7, $M_d = 9$. Distribution B: (0, 1, 1, 2, 2, 2, 3, 4, 5, 6, 6, 7, 8, 8), $N = 14$, M_d between cases 7 and 8, $M_d = (3+4)/2 = 3.5$. Distribution C: (1.6, 1.9, 1.9, 2.0, 2.0, 2.1, 2.1, 2.1, 2.4, 2.4, 2.5, 2.8), $N = 12$, M_d between cases 6 and 7, $M_d = (2.1 + 2.1)/2 = 2.1$.

Table 3-3. Means = 5, 8, 5, 5; medians = 5, 5, 5, 5; modes = none, none, 3 and 7, 5.

Table 3-4. $s_x^2 = 1$, $s_x = 1$, $s_y^2 = 9$, $s_y = 3$, $s_w^2 = 25$, $s_w = 5$.

Table 3-5. $\sum X_i = 36$, $\left(\sum X_i \right)^2 = 1296$, $N = 6$, $\sum X_i^2 = 232$, $\overline{X} = 6.0$, $M_d = 5.5$, $M_o = 5$, range $= 5$, $s^2 = 3.2$, $s = 1.79$.

Table 3-6. Depth of the median $= (70 + 1)/2 = 35.5$, $M_d = 8$; Depth of the fourths $= \dfrac{35 + 1}{2} = 18$; F_L is the 18th case from the bottom, $F_L = 11$; F_U is the 18th case from the top, $F_U = 25$; fourth-spread $= 25 - 11 = 14$; outliers exist below $11 - 1.5(14) = -10 = 0$ and above $25 + 1.5(14) = 46$, score value 47 is an outlier; $LEx = 0$, $UEx = 35$.

Table 3-7. $N = 70$, Measure = Looks at Mother, $M = 35.5$, $M_d = 18$, $F = 18$, $F_L = 11$, $F_U = 25$, $LEx = 0$, $UEx = 35$.

Self-Test. (1) The least squares sense: $\sum (X_i - \overline{X})^2$ is smaller than for any value other than $\overline{X}$. **(2)** The median is unaffected by extreme scores and therefore provides a more representative measure of central tendency when the distribution is substantially skewed. **(3)** The mean assumes an equal interval scale. **(4a)** The median is preferred because the distribution will be skewed to the right; **(4b)** The mode is preferred since the scale is nominal. (The proportion of cases in each category should also be reported.); **(4c)** The mean is preferred because height is usually unimodal and not skewed. (In a large distribution all three indices could be appropriate.) **(5)** The median and probably the mode. **(6)** It ignores the variability of scores between the smallest and largest values. **(7)** Since the standard deviation is expressed in original score units rather than squared units, it is easier to interpret. **(8)** Positive, because $\sum (X_i - \overline{X})^2$ and $N - 1$ will always be positive. **(9)** 0, 0. **(10)** A population is a collection of subjects, events, or scores that have some common characteristics of interest, and a sample is a subgroup of a population. **(11)** To make the statistic s^2 an unbiased estimator of σ^2. **(12)** Resistant indicators are less subject to the influence of unusual deviant scores. The stem-and-leaf display provides all the information in the frequency distribution, cumulative frequency distribution, and histogram in one display and preserves more of the detail of the data. The five-number summary and boxplot provide tabular and graphical representation to resistant indicators that convey control tendency, variability, outliers, and skewness.

Exercises. (1) For distribution X: $\overline{X} = 6.00$, $M_d = 6$, $M_o = $ no useful mode, range $= 6$, $s^2 = 4.67$, $s = 2.16$; for distribution Y: $\overline{Y} = 5.00$, $M_d = 5.00$, $M_o = 4.00$, range $= 7.00$, $s^2 = 5.33$, $s = 2.31$; for distribution Z: $\overline{Z} = 5.00$, $M_d = 5$,

$M_o = 5.00$, range $= 8$, $s^2 = 7.33$, $s = 2.71$. **(2a)** 3; **(2b)** 4; **(2c)** 5; **(2d)** 5; **(2e)** 25; **(2f)** 4.6. **(3)** See page 66 of text. **(4a)** $s^2 = 3.20$; **(4b)** $s^2 = 3.20$: this distribution is the same as in a except that a constant of -10 has been added to each score, so the variance remains the same; **(4c)** $s^2 = 3.20$: a

constant of 10 has been added to each score in a, so the variance remains the same; **(4d)** $s^2 = 320$: a constant of 10 has been multiplied with each score in a, so the variance is $10^2 = 100$ times that in a; see pages 102-104 of the text for a discussion of these relationships.

(5)

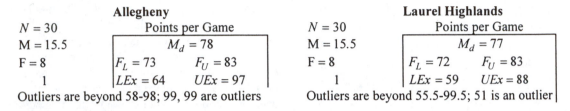

Allegheny		Laurel Highlands	
$N = 30$	Points per Game	$N = 30$	Points per Game
$M = 15.5$	$M_d = 78$	$M = 15.5$	$M_d = 77$
$F = 8$	$F_L = 73$ $F_U = 83$	$F = 8$	$F_L = 72$ $F_U = 83$
1	$LEx = 64$ $UEx = 97$	1	$LEx = 59$ $UEx = 88$

Outliers are beyond 58-98; 99, 99 are outliers   Outliers are beyond 55.5-99.5; 51 is an outlier

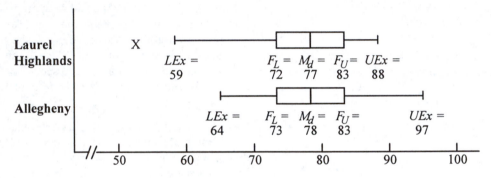

Generally, the teams are nearly the same except Allegheny can occasionally have very high-scoring games and Laurel Highlands can occasionally have very low-scoring games.

STATISTICAL PACKAGES

StataQuest
(To accompany the Guided Computational Example in Table 3-6.)

For a set of numbers, StataQuest will determine the count (Obs), mean, standard deviation (Std. Dev), variance, minimum, maximum, median (50%-tile), first quartile (25%-tile), and third quartile (75%-tile), and it will draw a boxplot.

Start the StataQuest program. To bring up the spreadsheet for data entry, point to the *File* menu and click *New*:

> *File>New*

or click on the *Editor* button at the top of the screen. The Stata Editor window should fill the screen.

This analysis will examine the number of looks at mother by a child given in Table 3-6. The *Looks* data consist of 70 observations, strung out in order as follows:

47	35	35	34	34	32	31	31	30	29	28	28	28	28	27	27	27	25	24	24
24	23	22	22	22	22	21	21	21	20	20	19	19	18	18	18	17	16	16	16
16	15	15	15	14	14	13	12	12	12	12	11	11	11	10	10	9	9	8	8
7	6	6	6	6	5	4	4	4	0										

Enter the number of looks recorded for each child in the first column of the spreadsheet. Label the first column with the variable name *Looks*. To enter the new variable name, point to the column heading and double click. The data entered should now appear on the spreadsheet as follows, in rows 1-70:

> Looks
> 47
> 35
> 35
> ⋮
> etc.

Save the *Looks* data that you entered:

> *File>Save as*
> (Enter drive and file name for the save file.)
> OK

Open a log file to save or print the results of the analysis. Click on the *Log* button at the top left of the screen, enter a name and drive, and click OK.

Two kinds of descriptive analyses can be carried out on the variable. The first strategy is to generate the usual frequency table and the mean and standard deviation:

Summaries>Tables>One-way (frequency)
        Data variable: Looks
        OK

Summaries>Means and SDs
        Data variable: Looks
        OK

The second strategy is to employ exploratory data analysis. In this section we will generate a table of descriptive statistics and a graphic boxplot display to complete the analysis. To get a table of descriptive statistics and a boxplot:

Summaries>Medians/Percentiles
        Data variable: Looks
        OK

Graphs>One variable>Boxplot
        Data variable: Looks
        OK

To print the output from all the above procedures:

File>Print Log
        OK
File>Print Graph
        OK

Exit StataQuest:

File>Exit

StataQuest Program Output

```
. tabulate Looks

    Looks|       Freq.      Percent        Cum.
---------+-----------------------------------
      0  |          1         1.43         1.43
      4  |          3         4.29         5.71
      5  |          1         1.43         7.14
      6  |          4         5.71        12.86
      7  |          1         1.43        14.29
      8  |          2         2.86        17.14
      9  |          2         2.86        20.00
     10  |          2         2.86        22.86
     11  |          3         4.29        27.14
     12  |          4         5.71        32.86
     13  |          1         1.43        34.29
     14  |          2         2.86        37.14
     15  |          3         4.29        41.43
     16  |          4         5.71        47.14
     17  |          1         1.43        48.57
     18  |          3         4.29        52.86
     19  |          2         2.86        55.71
     20  |          2         2.86        58.57
     21  |          3         4.29        62.86
     22  |          4         5.71        68.57
     23  |          1         1.43        70.00
     24  |          3         4.29        74.29
     25  |          1         1.43        75.71
     27  |          3         4.29        80.00
     28  |          4         5.71        85.71
     29  |          1         1.43        87.14
     30  |          1         1.43        88.57
     31  |          2         2.86        91.43
     32  |          1         1.43        92.86
     34  |          2         2.86        95.71
     35  |          2         2.86        98.57
     47  |          1         1.43       100.00
---------+-----------------------------------
   Total |         70       100.00
```

```
. summarize Looks

Variable |      Obs        Mean    Std. Dev.         Min         Max
---------+-----------------------------------------------------------
   Looks |       70    18.34286    9.561223           0          47


. summarize Looks, detail

                              Looks
-------------------------------------------------------------
      Percentiles     Smallest
 1%          0              0
 5%          4              4
10%          6              4         Obs                  70
25%         11              4         Sum of Wgt.          70

50%         18                        Mean           18.34286
                       Largest        Std. Dev.      9.561223
75%         25             34
90%         31             35         Variance       91.41698
95%         34             35         Skewness       .3761867
99%         47             47         Kurtosis       2.762814
```

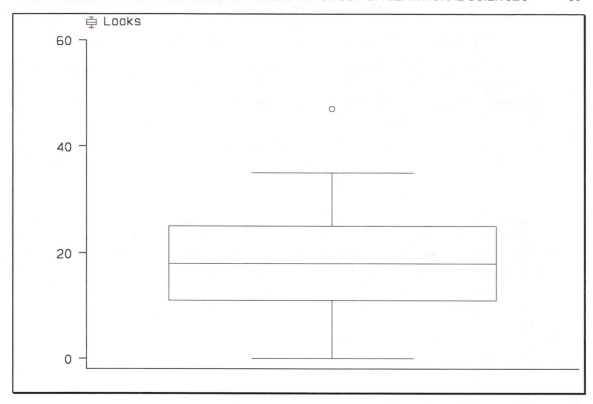

Notice that more information is provided than is covered in the text and that the boxplot is displayed vertically rather than horizontally as in the text and this *Guide*.

Minitab
(To accompany the guided Computational Example in Table 3-6.)

For a set of scores, Minitab will determine the mean, median, standard deviation (StDev), minimum, maximum, first quartile (Q1), and third quartile (Q3), and it will draw a boxplot.

This analysis will examine the number of looks at mother by a child given in Table 3-6. The *Looks* data ⌐ .sist of 70 observations, strung out in order as follows:

```
47  35  35  34  34  32  31  31  30  29  28  28  28  28  27  27  27  25  24  24
24  23  22  22  22  22  21  21  21  20  20  19  19  18  18  18  17  16  16  16
16  15  15  15  14  14  13  12  12  12  12  11  11  11  10  10   9   9   8   8
 7   6   6   6   6   5   4   4   4   0
```

Start the Minitab program. To enter data on the spreadsheet, click on the Data window. Enter the number of looks recorded for each child in the first column of the spreadsheet. Label the first column with the variable name *Looks*. To enter the new variable name, point to the column heading and double click. The data entered should now appear on the spreadsheet as follows, in rows 1-70:

> Looks
> 47
> 35
> 35
> ⋮
> etc.

Save the *Looks* data that you entered:

> *File>Save Worksheet As*
> (Enter drive and file name for the Minitab save file.)
> OK

Two kinds of analyses will be carried out on the variable. The first is the usual frequency table and the mean and standard deviation of the distribution. The second is exploratory data analysis, using the box-plot. Generate the frequency table with the Tally command and the statistics and boxplot with the Descriptive Statistics command:

> *Stat>Tables>Tally*
> Variables: Looks   [double-click to transfer]
> Display: Counts, Percents, Cumulative counts, Cumulative percents
> OK

> *Stat>Basic Statistics>Descriptive Statistics*
> Variables: Looks   [double-click to transfer]
> Graphs:
> > Boxplot of data: Yes
> > OK
> > OK

The above commands will yield tabular output in the Session window, and a graphic image of the box-plot. To print the results from the Session window:

> *Window>Session*
> *File>Print Window*
> > Print Range: All
> > OK

To print the boxplot:

> *Window>Boxplot of Looks*
> *File>Print Window*
> > Print Range: All
> > OK

Exit Minitab:

> *File>Exit*

Minitab Program Output

Summary Statistics for Discrete Variables

Looks	Count	CumCnt	Percent	CumPct
0	1	1	1.43	1.43
4	3	4	4.29	5.71
5	1	5	1.43	7.14
6	4	9	5.71	12.86
7	1	10	1.43	14.29
8	2	12	2.86	17.14
9	2	14	2.86	20.00
10	2	16	2.86	22.86
11	3	19	4.29	27.14
12	4	23	5.71	32.86
13	1	24	1.43	34.29
14	2	26	2.86	37.14
15	3	29	4.29	41.43
16	4	33	5.71	47.14
17	1	34	1.43	48.57
18	3	37	4.29	52.86
19	2	39	2.86	55.71
20	2	41	2.86	58.57
21	3	44	4.29	62.86
22	4	48	5.71	68.57
23	1	49	1.43	70.00
24	3	52	4.29	74.29
25	1	53	1.43	75.71
27	3	56	4.29	80.00
28	4	60	5.71	85.71
29	1	61	1.43	87.14
30	1	62	1.43	88.57
31	2	64	2.86	91.43
32	1	65	1.43	92.86
34	2	67	2.86	95.71
35	2	69	2.86	98.57
47	1	70	1.43	100.00
N=	70			

Descriptive Statistics

Variable	N	Mean	Median	Tr Mean	StDev	SE Mean
Looks	70	18.34	18.00	18.08	9.56	1.14

Variable	Min	Max	Q1	Q3
Looks	0.00	47.00	11.00	25.50

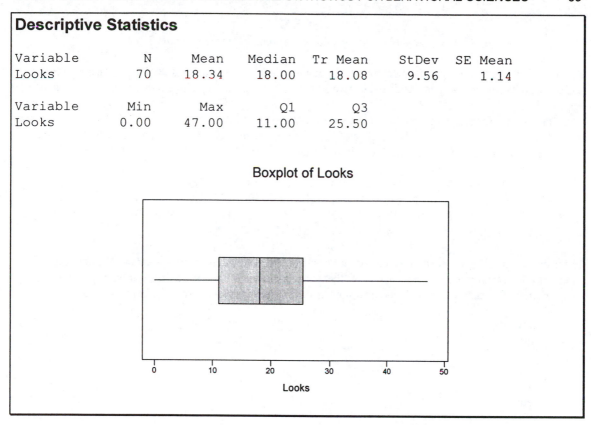

Boxplot of Looks

Notice that the line in the boxplot represents the range of scores, not the *LEx* to the *UEx*, so no outliers are plotted.

SPSS
(To accompany the guided computational example in Table 3-6.)

For a set of numbers, SPSS will determine the mean, median, mode, standard deviation, variance, range, and sum of scores. It also generates descriptive statistics for exploratory data analysis and produces a boxplot.

Start the SPSS program. The Data Editor window should fill the screen.

This analysis will examine the number of looks at mother by a child given in Table 3-6. The *Looks* data consist of 70 observations, strung out in order as follows:

```
47  35  35  34  34  32  31  31  30  29  28  28  28  28  27  27  27  25  24  24
24  23  22  22  22  22  21  21  21  20  20  19  19  18  18  18  17  16  16  16
16  15  15  15  14  14  13  12  12  12  12  11  11  11  10  10   9   9   8   8
 7   6   6   6   6   5   4   4   4   0
```

Enter the observations recorded for the 70 children in the first column of the spreadsheet. Label the column with the variable name *Looks*. To enter the new variable name, point to the column heading and double click. The data entered should now appear on the spreadsheet as follows, in rows 1-70:

> Looks
> 47
> 35
> 35
> ⋮
> etc.

Two kinds of descriptive analyses can be carried out on these data. The first strategy is to calculate a frequency table along with the traditional measures of central tendency and variability:

> *Statistics>Summarize>Frequencies*
> Variable(s): Looks
> Display frequency tables: Yes
> Statistics:
> > Central Tendency: Mean, Median, Mode, Sum
> > Dispersion: Std. Deviation, Variance, Range
> > Continue
> OK

The second strategy is to employ exploratory data analysis. A stem-and-leaf display can be generated in place of the frequency table, as was done in the previous chapter. In this section we will generate a table of descriptive statistics and a graphic display of a boxplot to complete the analysis:

> *Statistics>Summarize>Explore*
>> Dependent List: Looks
>> Display: Both
>> Plots:
>>> Boxplots: Factor levels together
>>> Descriptive: None
>>> Continue
>
> OK

To print the output generated by all the above procedures:

> *File>Print*
>> Print range: All visible output
>> OK

Save the *Looks* data that you entered:

> *Window>SPSS Data Editor*
> *File>Save as*
>> (Enter device and file name for SPSS save file.)
>> Save

Exit SPSS:

> *File>Exit SPSS*

SPSS Program Output

Frequencies

Statistics

	N		Mean	Median	Mode	Std. Deviation	Variance	Range	Sum
	Valid	Missing							
LOOKS	70	0	18.3429	18.0000	6.00[a]	9.5612	91.4170	47.00	1284.00

a. Multiple modes exist. The smallest value is shown

LOOKS

		Frequency	Percent	Valid Percent	Cumulative Percent
Valid	.00	1	1.4	1.4	1.4
	4.00	3	4.3	4.3	5.7
	5.00	1	1.4	1.4	7.1
	6.00	4	5.7	5.7	12.9
	7.00	1	1.4	1.4	14.3
	8.00	2	2.9	2.9	17.1
	9.00	2	2.9	2.9	20.0
	10.00	2	2.9	2.9	22.9
	11.00	3	4.3	4.3	27.1
	12.00	4	5.7	5.7	32.9
	13.00	1	1.4	1.4	34.3
	14.00	2	2.9	2.9	37.1
	15.00	3	4.3	4.3	41.4
	16.00	4	5.7	5.7	47.1
	17.00	1	1.4	1.4	48.6
	18.00	3	4.3	4.3	52.9
	19.00	2	2.9	2.9	55.7
	20.00	2	2.9	2.9	58.6
	21.00	3	4.3	4.3	62.9
	22.00	4	5.7	5.7	68.6
	23.00	1	1.4	1.4	70.0
	24.00	3	4.3	4.3	74.3
	25.00	1	1.4	1.4	75.7
	27.00	3	4.3	4.3	80.0
	28.00	4	5.7	5.7	85.7
	29.00	1	1.4	1.4	87.1
	30.00	1	1.4	1.4	88.6
	31.00	2	2.9	2.9	91.4
	32.00	1	1.4	1.4	92.9
	34.00	2	2.9	2.9	95.7
	35.00	2	2.9	2.9	98.6
	47.00	1	1.4	1.4	100.0
	Total	70	100.0	100.0	
Total		70	100.0		

Explore

Descriptives

			Statistic	Std. Error
LOOKS	Mean		18.3429	1.1428
	95% Confidence Interval for Mean	Lower Bound	16.0631	
		Upper Bound	20.6227	
	5% Trimmed Mean		18.0952	
	Median		18.0000	
	Variance		91.417	
	Std. Deviation		9.5612	
	Minimum		.00	
	Maximum		47.00	
	Range		47.00	
	Interquartile Range		14.5000	
	Skewness		.384	.287
	Kurtosis		-.164	.566

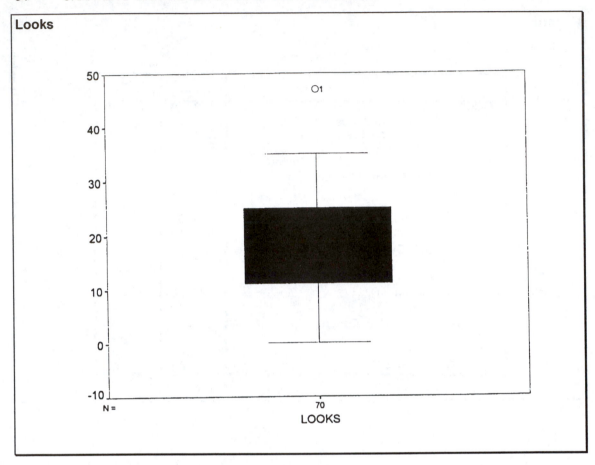

Notice that substantially more information is provided than has been presented thus far in the text. Further, the boxplot is displayed vertically, rather than horizontally.

CHAPTER 4

INDICATORS OF RELATIVE STANDING

CONCEPT GOALS

Be sure that you thoroughly understand the following concepts and how to use them in statistical applications.

- ◆ Percentile points and ranks
- ◆ Standard scores, z
- ◆ Mean, variance, and standard deviation of standard scores
- ◆ Normal distribution
- ◆ Standard normal distribution

GUIDE TO MAJOR CONCEPTS

Percentiles

Simply knowing that you scored 88 on a statistics exam does not
always tell you much about how well you did. For example, you
would feel somewhat better if you knew that 88 was above the mean,
but even if you knew that 88 was 10 points above the mean, you
might still wonder how many other students scored that well. Thus,
we need techniques that indicate the place of individual scores in a
distribution. These techniques are called indicators of

_____ _____ .   relative standing

The most common indicators of relative standing are the
percentile point and **percentile rank**. The score value below which
the proportion *P* of the cases in the distribution lie is the *P*th

_____ _____ . The proportion *P* of the cases is known   percentile point
as the _____ _____ of that score value. For example,   percentile rank
if .75 of the cases in a distribution scored lower than 88, then 88 is
the _____ _____ and .75 is its _____   percentile point; percentile
_____ . Percentile points are score values, but percentile ranks   rank
are proportions, designated by a capital *P* with a subscript
corresponding to the desired proportion. Therefore, the 75th
percentile rank is symbolized by _____ . Since proportion times 100 $P_{.75}$
equals percent, P_{75}, called the 75th _____ _____ ,   percentile rank
refers to the score value, or _____ _____ , for   percentile point
which ____ % of cases lie (above/below) _____ its value. As   75; below
such, percentile rank is identical to the _____ _____   cumulative relative
_____ of its corresponding score value.   frequency

In the last chapter, we calculated the median for a group of
scores. Since the median is that point in the distribution below which
.50 of the cases fall, the median is identical to the percentile rank
_____ . In addition, we might want to know how to determine $P_{.50}$
percentile points corresponding to percentile ranks other than $P_{.50}$.
The percentile point corresponding to the *P*th percentile rank is given
by

$$Percentile\ point\ X = L + \left[\frac{P(N) - n_b}{n_w} \right] i$$

in which

$P =$ the percentile rank of the required point (that is, the
proportion of cases falling below the desired point; *P*
ranges between 0 and 1.00)

$L =$ the lower real limit of the score value containing the required percentile point

$N =$ the total number of cases in the entire distribution

$n_b =$ the number of cases falling below L

$n_w =$ the number of cases falling within the score value containing the required percentile point

$i =$ the size of the score value measurement unit ($i = 1$ if the data are in whole numbers; $i = .1$ if the data are in tenths; and so on)

This formula looks more forbidding than it really is. Notice that $P(N)$ is simply the number of cases in the distribution which must lie below the desired score value. Suppose that the distribution at hand is the one presented in Table 4-1. Then if we want the 35th percentile point and there are 40 observations in the sample, $P =$ ___ , $N =$ ___ , and $P(N) =$ _____ = ____ . Therefore, we seek the score value such that ____ cases fall (above/below) _____ that point. Looking at Table 4-1, the fourteenth lowest score is ____ . Since the score value 61 represents the interval from 60.5 to 61.5, this is the interval which must contain the desired percentile point. Therefore, $L =$ ___ , $P(N) =$ ___ , $n_b =$ ___ , $n_w =$ ___ and $i =$ ___ , and

$$X = \underline{\hspace{3cm}}$$

$$= \underline{\hspace{4cm}} = \underline{\hspace{3cm}}$$

.35

$40; .35(40) = 14$

14; below

61

60.5; 14; 13; 1; 1

$$L + \left[\frac{P(N) - n_b}{n_w} \right] i$$

$$= 60.5 + \left[\frac{14 - 13}{1} \right] 1 = 61.5$$

Therefore, 61.5—the upper real limit of the 14th score—is the score value below which 35% of the distribution falls. In symbols,

$$\underline{\hspace{2cm}} = \underline{\hspace{1.5cm}} .$$

Similarly, if we want to compute the 45th percentile point for the data in Table 4-1, first calculate $P(N) =$ _____ = ____ to determine that the desired point lies in the interval _____ . Then $L =$ ___ , $P(N) =$ ___ , $n_b =$ ___ , $n_w =$ ___ , $i =$ ___ , and

$$X = \underline{\hspace{3cm}}$$

$$= \underline{\hspace{2.5cm}} = \underline{\hspace{2.5cm}}$$

So, _____ = ____ .

$P_{35} = 61.5$

$.45(40) = 18$

71.5 to 72.5

71.5; 18; 17; 3; 1

$$L + \left[\frac{P(N) - n_b}{n_w} \right] i$$

$$= 71.5 + \left[\frac{18 - 17}{3} \right] 1 = 71.83$$

$P_{45} = 72$

Table 4-1 Distribution of Scores on a Statistics Examination for a Class of Forty Students

Student	Score	Student	Score	Student	Score	Student	Score
40	97	30	80	20	72	10	56
39	93	29	79	19	72	9	55
38	92	28	78	18	72	8	55
37	91	27	78	17	71	7	55
36	88	26	78	16	70	6	55
35	85	25	78	15	65	5	51
34	85	24	76	14	61	4	49
33	84	23	75	13	60	3	49
32	83	22	75	12	59	2	48
31	82	21	74	11	58	1	46

Sometimes we know the score and want to find its percentile rank. The formula above can be transposed to give

$$P = \frac{n_w(X - L) + in_b}{Ni}$$

in which

$P =$ the desired percentile rank (that is, the proportion of cases lying below X; P ranges between 0 and 1.00)

$X =$ the score value for which the percentile rank is desired

$L =$ the lower real limit of X

$n_w =$ the number of cases having the score value X

$n_b =$ the number of cases having a score value lower than X

$N =$ the total number of cases in the entire distribution

$i =$ the size of the score value measurement unit ($i = 1$ if the data are in whole numbers; $i = .1$ if the data are in tenths; and so on)

Looking at Table 4-1, $X =$ ___ , $L =$ ___ , $n_w =$ ___ , $n_b =$ ___ ,

$N =$ ___ , $i =$ ___ , and

$$P = \underline{\hspace{3cm}}$$

$$= \underline{\hspace{3cm}} = \underline{\hspace{2cm}}$$

In symbols, ___ = ___ .

74; 73.5; 1; 20

40; 1

$$\frac{n_w(X - L) + in_b}{N_i}$$

$$= \frac{1(74 - 73.5) + 1(20)}{40(1)} = .5125$$

$$P_{51} = 74$$

Suppose that the percentile rank for the score 55 is desired.
Then, $X =$ ___ , $L =$ ___ , $n_w =$ ___ , $n_b =$ ___ , $N =$ ___ , $i =$ _____ ,

55; 54.5; 4; 5; 40; 1

and

$$P = \underline{\hspace{3cm}}$$

$$\frac{n_w(X - L) + in_b}{N_i}$$

$$= \underline{\hspace{4cm}} = \underline{\hspace{3cm}}$$

$$= \frac{4(55 - 54.5) + 1(15)}{40(1)} = .175$$

In symbols, ____ = ____ .

$$P_{.175} = 55$$

Standard Scores

In statistics we often convert scores on one scale (e.g., the X scale)
into scores on another scale, called z, by the formula

$$z_i = \frac{X_i - \overline{X}}{s_x}$$

Such a z_i is called a **standard score**. This transformation to a
_____ _____ is common in the social and behavioral

standard score

sciences. The distribution of X's with a mean $\overline{X}$ and a standard
deviation s_x becomes a distribution of z's which has mean 0 and
standard deviation 1.00. To illustrate this, consider the distribution of
X's presented in Table 4-2, which has a mean $\overline{X} = 5$ and a standard
deviation $s_x = 2$. Transform the X's into z's with the above formula,
and fill in the columns of the table. Then calculate the mean and
standard deviation of the z's. You should find that the mean of the z's
is $\overline{z} =$ ___ and that the standard deviation of the z's is $s_z =$ ___ .

0; 1.00

Now look at the bottom of the table. The X's have been located
on the X scale of measurement with boxes. Position the z scores in
their proper place on the z scale in the same way. Notice that the
relative position of the scores in the distribution of z's (is/is not) ___

is

the same as for the X's. Standardizing does not change the form of
the distribution, only its _____ of measurement.

scale

The fact that the standard deviation of the z's is always equal to
_____ implies that the unit of measurement of the z scale is equal

1.00

to its _____ _____ . Therefore, $z = 1.50$ indicates

standard deviation

that such a score falls 1.5 _____ _____ to the right of

standard deviations

the mean. Also, in an X distribution with mean 50 and standard
deviation 10, a score of 70 is ___ standard deviations above the

2

mean. Thus, its corresponding z score is ___ .

2.00

In summary, the transformation of any X distribution to z units
employs the formula

$$z_i = \underline{\hspace{3cm}}$$

$$\frac{X_i - \overline{X}}{s_x}$$

Table 4-2 Converting X_i to z_i

X_i	$\overline{X}$	$z_i = \dfrac{X_i - \overline{X}}{s_x}$	$(z_i - \overline{z})$	$(z_i - \overline{z})^2$
2	5			
4	5			
5	5			
5	5			
6	5			
8	5			

$$\sum X_i = 30 \qquad\qquad \sum z_i = \qquad\qquad \sum (z_i - \overline{z})^2 =$$

$$N = 6 \qquad\qquad N = \qquad\qquad N - 1 =$$

$$\overline{X} = 5 \qquad\qquad \overline{z} = \frac{\sum z_i}{N} = \qquad\qquad s_z^2 = \frac{\sum (z_i - \overline{z})^2}{N - 1} =$$

$$s_x = 2 \qquad\qquad\qquad\qquad s_z = \sqrt{s_z^2} =$$

X: scale from 0 to 8

z: scale from -2.5 to 1.5

Although the mean and standard deviation of the original X distribution were respectively $\overline{X}$ and s_x, the mean of the z's will always be ___ and the standard deviation will always be ___ , **0; 1** regardless of the values of $\overline{X}$ and s_x. Thus, z scores are expressed in _____ _____ units. It is in this sense that z **standard deviation** scores are said to be _____ _____ , because any X **standard scores** distribution can be transformed into a z distribution with $\overline{z} = $ ___ and **0** $s_z = $ ___ . **1.00**

Standard Normal Distribution

Converting to z scores has distinct advantages if the original X distribution is a special type, one having a frequency distribution that looks something like Figure 4-1, which statisticians call a _____ **normal** distribution. A normal distribution is defined theoretically by a specific mathematical equation, and there are normal distributions having different means and standard deviations. However, if the

original X distribution is normal in form and then the scores are
converted to z scores, such a distribution of z scores is called the
standard _____ distribution.   normal

     Although any distribution can be transformed into a z
distribution, not any distribution can be transformed into a standard
normal distribution; this is possible only if the original X distribution
is a _____ distribution. Normality of the X distribution is   normal
required because transforming to z scores does not change the form
of the distribution (look back at your work in Table 4-2.)

     An important feature of the standard normal distribution is that
all the percentile ranks are known for each possible value of z. In
fact, the percentiles of the _____ _____ distribution   standard normal
are presented in Table A in Appendix 2 of the text. Open your text to
Table A. The column labeled z lists z values from 0.00 to 4.00; the
other two columns list the proportion of the area under the normal
curve that lies between certain z values. We can interpret proportion
of area as proportion of cases, so simply remember the verbal
equation: proportion of _____ = proportion of _____ . For each z area; cases
in Table A, the proportion of cases found between the mean of the z
distribution ($\bar{z} = 0$) and the listed value of z is given in the second
column; the proportion of cases having higher z scores than the listed
z value is presented in the third column. To find the percentage of
cases, note that the percentage = 100 times the proportion. For
example, for a z value of 1.25, the second column of the table states
that .3944, or 39.44%, of the cases in a standard normal distribution
would have z values between $z =$ ___ and $z =$ ___ . The third column   0; 1.25
indicates that ___ , or _____ %, would have z scores (lower/higher)   .1056; 10.56
_____ than 1.25.   higher

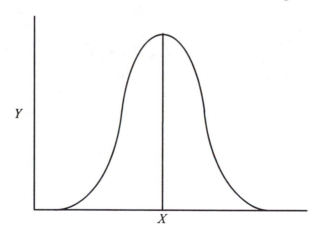

Fig. 4-1. A normal distribution

You can use Table A to obtain percentile ranks and other information for any z value in a normal distribution if you remember a few facts about the standard normal distribution. First, the standard normal is a symmetrical distribution with one mode. Therefore, the mean, median, and mode all have the same value, namely $z =$ ____ .

Second, z values to the left of 0 are _____ , while those to the right of $z = 0$ are ____ . Third, since the distribution is symmetrical about 0, the proportion of cases between $z = 0$ and $z = +.75$ is equal to the proportion of cases between $z = 0$ and $z =$ ___ .

Given this information, a number of problems can be solved with Table A. First, the proportion of cases falling to the right of $z = +.61$, for example, is approximately _____ , and the proportion falling to the left of $z = -.61$ is _____ .

The proportion of cases falling to the *left* of $z = +.61$ is the sum of the proportion between $z = 0$ and $z = .61$ which is _____ plus the proportion to the left of $z = 0$ which is ____ for a total of _____ .

The proportion of cases falling between $z = -1.00$ and $z = +1.00$ can also be determined from Table A. Problems of this sort can be solved more easily if you sketch the given information on a standard normal distribution. In the first blank graph of Figure 4-2, draw vertical lines through the distribution at $z = -1.00$ and $z = +1.00$, and shade in the area between them. You can now see that the proportion of area between $z = -1.00$ and $+1.00$ is the amount between $z = 0$ and $z = +1.00$ plus the amount between $z = 0$ and $z = -1.00$. Because the normal curve is _____ about $z = 0$, the required proportion is ____+____=____ . Using Table A and the graphs provided, determine the proportion of area between and $z = -1.96$ and $z = +1.96$: _____+_____=_____ ; between $z = 1.73$ and $z = 1.02$:_____-_____=_____ ; and between $z = -1.56$ and $z = -.21$: _____-_____=_____ .

Now suppose that an X distribution is normal, with $\overline{X} = 70$ and $s_x = 5$. What proportion of the cases scored higher than $X = 77$? To solve this problem, $X = 77$ first must be translated into a z score as follows:

$$z = \frac{X_i - \overline{X}}{s_x} = \underline{\hspace{2cm}} = \underline{\hspace{2cm}}$$

Then, by looking in Table A, determine the proportion of cases that lies to the (right/left) _____ of $z =$ ___ , which is ___ .

0
negative
positive

$-.75$

.27
.27

.2291
.5000; .7291

symmetrical
$.3413+.3413=.6826$

$.4750+.4750=.9500$
$.4582-.3461=.1121$
$.4406-.0832=.3574$

$\dfrac{77-70}{5}=1.40$

right; 1.40; .0808

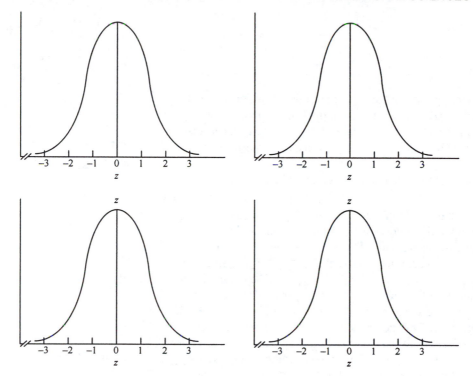

Fig. 4-2. Standard normal distributions

What proportion of cases fall between $X = 95$ and $X = 88$ in a normal distribution with $\overline{X} = 85$ and $s_x = 5$? First, look at Figure 4-3, which presents this normal distribution. Locate the points 88 and 95 on the X scale, draw vertical lines at these points, and shade in the area between them. This is the area requested. Now, transform these X values into their z equivalents:

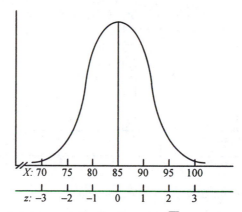

Fig. 4-3. Normal distribution with $\overline{X} = 85$ and $s_x = 5$

For $X = 95: z =$ _____ $=$ _____ $=$ _____

$$\frac{X_i - \overline{X}}{s_x} = \frac{95 - 85}{5} = 2.00$$

For $X = 88: z =$ _____ $=$ _____ $=$ _____

$$\frac{X_i - \overline{X}}{s_x} = \frac{88 - 85}{5} = .60$$

Mark these points on the z scale in Figure 4-3. The required area can be determined by obtaining the proportion of area between $z = 0$ and $z = 2.00$, which is _____ , and subtracting the area between $z = 0$ and $z = .60$, which is _____ . Thus, the required area is _____ $-$ _____ $=$ _____ .

.4772
.2257
$.4772 - .2257 = .2515$

Suppose that you are asked a slightly different question: In a normal distribution with $\overline{X} = 40$ and $s_x = 10$, what score value has a percentile rank of .67? Look at Figure 4-4, which displays the standard normal distribution. First, determine what point along the z scale is at the 67th percentile. This is the z value for which _____ % of the total area lies to its (left/right) _____ . Since the proportion of area to the left of $z = 0$ is _____ , the required point must be such that $.67 - .50 =$ _____ of the area falls between $z = 0$ and this point. These values are already sketched in on Figure 4-4. Go to Table A and look down the *second* column, which gives the proportion of area between $z = 0$ and the z of that row of the table, until you find the value _____ . The corresponding z value is $z =$ _____ . Mark this on the z scale in Figure 4-4. Translating this z into an X_i score in a distribution with $\overline{X} = 40$ and $s_x = 10$ can be accomplished with the usual formula and a little algebra:

67; left
.5000
.17

.1700; .44

$$z = \frac{X_i - \overline{X}}{s_x}$$

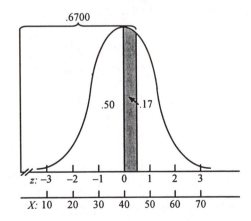

Fig. 4-4. Determining the 67th percentile point in the standard normal distribution

Substitute the known values of $z_i = .44$, $\overline{X} = 40$, and $s_x = 10$, and solve for X_i:

_____	=	_____	$.44 = \dfrac{X_i - 40}{10}$
_____	=	_____	$.44 = \dfrac{X_i}{10} - 4$
_____	=	_____	$4.44 = \dfrac{X_i}{10}$
$X_i =$ _____			44.4

Therefore, the score value with a percentile rank of 67 is _____ . 44.4

Steps in solving these kinds of problems are summarized in Table 4-3.

Table 4-3 Steps in Determining Proportions of Cases in Normal Distribution

1. Draw a picture of a normal distribution.
2. Put the X scale under the abscissa if X values are given in the problem; put the z scale under the abscissa if z values are given in the problem.
3. Put the other scale (z or X) as an extra abscissa under the previous scale in the manner of Figures 4-3 and 4-4.
4. Locate on the appropriate scale any X or z values given in the problem.
5. Shade in the area under the curve requested by or given in the problem.
6. Convert relevant X scores to z scores or the reverse.
7. Determine relevant areas or z values from Table A, Appendix 2 of the text.
8. Use the above information to determine the specific answer to the question (eg., add or subtract appropriate areas.)

SELF TEST

1. In any distribution, if a score of 50 is at the 75th percentile, 75 is the percentile _____ corresponding to the percentile _____ of 50.

2. Your score is at the 72nd percentile rank in a sample of 75 people. How many people scored above you?

3. If you earn a score of 97 on a mathematics exam and a score of 23 on an English exam, is it possible for your percentile rank to be higher in English than in mathematics? Explain.

4. Suppose that your score of 23 on the English exam gives you a percentile rank of 85. One week after returning your exams, your instructor announces that an arithmetic error has been discovered and that he is increasing all exam scores by 30 points. What will be your new percentile rank?

5. What is the relationship between cumulative relative frequency and percentile rank?

*6. In the equation $Y = kX + c$ in which X is transformed to Y,
   a. Which term(s) are constants?

b. Which term(s) change the location of the origin?

c. Which term(s) change the size of the unit of measurement?

d. Which term(s) change the variance of the distribution?

*7. If we have an X distribution such that $\overline{X} = 20$ and $s_x = 4$, and if X is then transformed into Y by the equation $Y = 3X + 1$, what will $\overline{Y}$, $s_y{}^2$, and s_y equal?

*8. Write the transformation equation and answer the same questions as in question 7 with the X's transformed into the z distribution.

*9. Why is it advantageous for a personnel selection officer to convert into standardized form the scores of applicants on three different selection tests, rather than just averaging the particular scores before deciding which applicant to hire?

10. In the standard normal distribution, the mean is _____ , the variance is _____ , the standard deviation is _ _____ , and the total area under the curve is _____ .

11. Suppose there are two tests that are equally valid in predicting whether a trainee will successfully complete a particular program.

Candidate A scores 22 on the first test and 102 on the second test, for a total of 124. Candidate B scores 20 on the first test and 104 on the second test, also for a total of 124. Given that the first test has a mean of 20 and standard deviation of 2, and that the second test has a mean of 100 and a standard deviation of 20, which candidate appears more likely to be successful in the training program?

12. If John scored 23 on a geology test and this was equivalent to a z score of .60, what was the standard deviation of the scores in John's class if the mean was 20?

*13. If the mean of a normal distribution of history exam grades is 80 and the standard deviation is 5, all (or essentially all, that is, 99.74%) the scores will probably fall between what two values?

*14. Suppose that you have a sample of 25 scores on a test of scholastic performance and you want to use the standard normal distribution to determine percentiles for each student. What two assumptions must you make to do this? Explain why.

*15. If scores on a test of mathematical ability are transformed by the equation $Y = 3\left[(X_i - \overline{X})/s_x\right] + 5$, what will be the mean and standard deviation of the Y distribution?

*Questions preceded by an asterisk can be answered on the basis of the discussion in the text, but the discussion in this Study Guide does not answer them.

EXERCISES

Exercises 1-3 refer to the following data.

Subject	X	Subject	X	Subject	X	Subject	X
40	95	30	88	20	83	10	73
39	94	29	87	19	83	9	72
38	94	28	86	18	82	8	71
37	92	27	86	17	80	7	71
36	91	26	86	16	80	6	68
35	89	25	85	15	79	5	63
34	88	24	85	14	78	4	57
33	88	23	84	13	75	3	53
32	88	22	84	12	75	2	53
31	88	21	84	11	73	1	51

1. Determine the percentile points corresponding to the following ranks.
   a. $P_{.10}$ e. $P_{.75}$
   b. $P_{.225}$ f. $P_{.25}$
   c. $P_{.60}$ *g. The interquartile range
   d. $P_{.80}$

2. Determine percentile ranks for the following score values.
   a. 72   c. 84   e. 88
   b. 80   d. 86   f. 95

3. In what sense is it correct and in what sense is it incorrect to say that someone who scored at $P_{.75}$ did three times as well as someone who scored at $P_{.25}$ on the same test?

*4. Given an X distribution having $\overline{X} = 100$ and $s_x = 10$, determine the mean, variance, and standard deviation after the following transformations.

Transformation	Mean	Variance	Standard Deviation
none	100	100	10
$Y = X + 10$			
$Y = 2X$			
$Y = 3X + 4$			
$Y = 4(X + 1)$			

5. Determine the z score for X_i under the following conditions.

$\overline{X}$	s_x	X_i	z
35	5	45	
25	3	23.5	
50.3	3.3	49.2	
−15	6.5	−7.7	

6. Determine the proportion of cases in the standard normal distribution that fall between the mean and a z score of
   a. −1.00   c. .55   e. −3.17
   b. 1.96   d. 1.33   f. 0

7. Determine the proportion of cases in the standard normal distribution which have z scores higher than
   a. −1.00   c. .75
   b. 1.96   d. 0

8. Determine the percentile rank of the following z scores in the standard normal distribution.
   a. 1.96   c. 0   e. .04
   b. −.35   d. 2.33   f. .35

9. Determine the proportion of cases in the standard normal distribution falling between the following z values.
    a. −1.96 and +1.96   d. −.50 and +1.50
    b. −1.645 and +1.645 e. −1.33 and −.33
    c. −1.00 and +1.00   f. .45 and 1.20

10. In a normal population with mean 20 and standard deviation 5, what proportion of cases score
    a. more than 15? 18? 20? 21? 30?
    b. between 15 and 25? 10 and 30? 17.2 and 21.5? 21.5 and 22?

11. In the distribution in question 10, what score did a person have if $z = .43$? If $z = −.54$? If that person was at $P_{.50}$? At $P_{.33}$?

12. Assume that scores on the Scholastic Assessment Test are normally distributed with mean 500 and standard deviation 100.
    a. If Snoot U. takes students only from the top 20%, what score should you have before you consider applying?
    b. If Smithville College will consider anyone with a score of 600 or better, what percentage fewer applicants can Smithville expect than Snoot U.?
    c. What score does it take to be in the top 4%?

ANSWERS

Self-Test. (1) Rank, point. **(2)** 21. **(3)** Yes, because percentile ranks are relative positions, and all the scores on your mathematics test could be higher than all the scores on the English exam. **(4)** You remain at the 85th percentile. **(5)** Cumulative relative frequency and percentile rank are essentially the same thing. **(6a)** k and c; **(6b)** c; **(6c)** k; **(6d)** k. **(7)** 61; 144; 12. **(8)** $z = (X_i − \overline{X})/s_x$; 0; 1; 1. **(9)** The results of three different tests can be compared more accurately if they are converted to distributions, such as the z, which will have the same mean and standard deviation. **(10)** 0; 1; 1; 1.00. **(11)** Candidate A. **(12)** 5. **(13)** 65 and 95. **(14)** You must assume that the population distribution of scores on the test is normal in form and that the sample is large enough for $\overline{X}$ and s_x to be reasonably accurate estimators of the mean and standard deviation of that population distribution. **(15)** Mean = 5; standard deviation = 3.

Exercises. (1a) 57.5; **(1b)** 72.5; **(1c)** 85.0; **(1d)** 88.1; **(1e)** 87.70; **(1f)** 73.0; **(1g)** 73.0 to 87.70. **(2a)** $P_{.2125}$; **(2b)** $P_{.40}$; **(2c)** $P_{.5375}$; **(2d)** $P_{.6625}$; **(2e)** $P_{.7875}$; **(2f)** $P_{.9875}$. **(3)** Three times as many people scored below $P_{.75}$ as below $P_{.25}$, but the numerical value of the percentile point corresponding to $P_{.75}$ would probably not be three times that of $P_{.25}$. **(4)** Means = 110, 200, 304, 404; variances = 100, 400, 900, 1600; standard deviations = 10, 20, 30, 40. **(5)** 2.00; −.50; −.33; 1.12. **(6a)** .3413; **(6b)** .4750; **(6c)** .2088; **(6d)** .4082; **(6e)** .4992; **(6f)** 0. **(7a)** .8413; **(7b)** .0250; **(7c)** .2266; **(7d)** .5000. **(8a)** $P_{.9750}$; **(8b)** $P_{.3632}$; **(8c)** $P_{.5000}$; **(8d)** $P_{.9901}$; **(8e)** $P_{.5160}$; **(8f)** $P_{.6368}$. **(9a)** .9500; **(9b)** .9000; **(9c)** .6826; **(9d)** .6247; **(9e)** .2789; **(9f)** .2113. **(10a)** .8413; .6554; .5000; .4207; .0228. **(10b)** .6826; .9544; .3302; .0375. **(11)** 22.15; 17.3; 20.0; 17.8. **(12a)** 584 to 585; **(12b)** 4.13%; **(12c)** 675.

STATISTICAL PACKAGES

StataQuest
(To accompany Table 4-1.)

StataQuest can generate frequency and cumulative relative frequency distributions for raw scores, from which certain percentiles can be determined. However, like most packages, StataQuest will not calculate percentile ranks or points on request and does not interpolate when duplication of scores exists near a desired percentile, so answers may differ slightly from the sample answers in the *Guide*. To convert raw scores to *z* scores, a new variable must be calculated. This requires obtaining the mean and standard deviation of the distribution and entering them into the formula for the *z* score.

Start the StataQuest program. The data in Table 4-1 are:

97 93 92 91 88 85 85 84 83 82 80 79 78 78 78 78 76 75 75 74 72 72 72 71 70 65 61 60 59 58 56 55 55 55 55 51 49 49 48 46

Bring up the Stata Editor window and enter the 40 values for the variable *Scores* in the first column. Label the column with the variable name. To enter the variable name, point to the column heading and double click. The data entered should appear on the spreadsheet as follows, in rows 1-40:

Scores
97
93
92
⋮
etc.

Save the *Scores* data that you entered:

File>Save as
        (Enter drive and file name for the save file.)
        OK

Open a log file to save or print the results of the analysis. Click on the *Log* button at the top left of the screen, enter a name and drive, and click OK.

To generate standard scores, first determine the mean and standard deviation of the *Scores* variable:

Summaries>Means and SDs
        Data variables: Scores   [click to transfer]
        OK

Then calculate a new variable to contain the *z* scores. Type the values you determined for the mean [70.75] and the standard deviation [14.20139] into the formula for *z*, which is (Raw Score − Mean) / (Standard Deviation):

Data>Generate>Replace>Formula
    New variable name: **Zscores**   [type in]
    Formula: **(Scores-70.75)/14.20139**   [type in the formula]
    OK

To generate the frequency table for the raw scores:

Summaries>Tables>One-way (frequency)
    Data variable: Scores   [click to transfer]
    OK

To generate the frequency table for the standard scores:

Summaries>Tables>One-way (frequency)
    Data variable: Zscores   [click to transfer]
    OK

Note that only intervals with one or more counts are displayed in the frequency tables that were output. The interval around the raw score of 47, for instance, is not displayed in the table because it had no frequencies. To determine the percentile rank for a raw score of 49, look at the cumulative percent ("Cum.") for the score value of 49, which is 10.0, indicating that a score of 49 is at the 10th percentile rank. Percentile points can be determined only for those percentiles listed in the table under "Cum." Remember, the program does not interpolate.
    To print the output generated by the previous commands:
        File>Print Log
            OK

Exit StataQuest:

    File>Exit

StataQuest Program Output

```
. summarize Scores

Variable |      Obs        Mean    Std. Dev.        Min        Max
---------+-----------------------------------------------------------
  Scores |       40       70.75    14.20139         46         97


. tabulate Scores

     Scores|      Freq.      Percent        Cum.
-----------+-----------------------------------
        46 |         1         2.50        2.50
        48 |         1         2.50        5.00
        49 |         2         5.00       10.00
        51 |         1         2.50       12.50
        55 |         4        10.00       22.50
        56 |         1         2.50       25.00
        58 |         1         2.50       27.50
        59 |         1         2.50       30.00
        60 |         1         2.50       32.50
        61 |         1         2.50       35.00
        65 |         1         2.50       37.50
        70 |         1         2.50       40.00
        71 |         1         2.50       42.50
        72 |         3         7.50       50.00
        74 |         1         2.50       52.50
        75 |         2         5.00       57.50
        76 |         1         2.50       60.00
        78 |         4        10.00       70.00
        79 |         1         2.50       72.50
        80 |         1         2.50       75.00
        82 |         1         2.50       77.50
        83 |         1         2.50       80.00
        84 |         1         2.50       82.50
        85 |         2         5.00       87.50
        88 |         1         2.50       90.00
        91 |         1         2.50       92.50
        92 |         1         2.50       95.00
        93 |         1         2.50       97.50
        97 |         1         2.50      100.00
-----------+-----------------------------------
     Total |        40       100.00
```

```
. replace Zscores = (Scores-70.75)/14.20139
(40 real changes made)


. tabulate Zscores

    Zscores|      Freq.      Percent        Cum.
------------+-----------------------------------
-1.742787 |          1         2.50         2.50
-1.601956 |          1         2.50         5.00
 -1.53154 |          2         5.00        10.00
-1.390709 |          1         2.50        12.50
-1.109046 |          4        10.00        22.50
-1.038631 |          1         2.50        25.00
-.8977994 |          1         2.50        27.50
-.8273838 |          1         2.50        30.00
-.7569681 |          1         2.50        32.50
-.6865525 |          1         2.50        35.00
-.4048899 |          1         2.50        37.50
-.0528117 |          1         2.50        40.00
 .0176039 |          1         2.50        42.50
 .0880195 |          3         7.50        50.00
 .2288508 |          1         2.50        52.50
 .2992665 |          2         5.00        57.50
 .3696821 |          1         2.50        60.00
 .5105134 |          4        10.00        70.00
  .580929 |          1         2.50        72.50
 .6513447 |          1         2.50        75.00
 .7921759 |          1         2.50        77.50
 .8625916 |          1         2.50        80.00
 .9330072 |          1         2.50        82.50
 1.003423 |          2         5.00        87.50
  1.21467 |          1         2.50        90.00
 1.425917 |          1         2.50        92.50
 1.496332 |          1         2.50        95.00
 1.566748 |          1         2.50        97.50
 1.848411 |          1         2.50       100.00
------------+-----------------------------------
    Total |         40       100.00
```

Minitab
(To accompany Table 4-1.)

Minitab will produce percentiles and calculate standard scores for all values in a distribution.
  Start the Minitab program. The data in Table 4-1 are:

97 93 92 91 88 85 85 84 83 82 80 79 78 78 78 78 76 75 75 74 72 72 72 71 70 65 61 60
59 58 56 55 55 55 55 51 49 49 48 46

  To enter data on the spreadsheet, click on the Data window. Enter the 40 score values in the first column. Label the top of the column with the variable name *Scores*. To enter the variable name, click on the space at the top of the column and type in the name. The data you entered should appear on the spreadsheet as follows, in rows 1-40:

    Scores
    97
    93
    92
    ⋮
    etc.

Save the *Scores* data that you entered:

    File>Save Worksheet As
      (Enter drive and file name for the Minitab save file.)
      OK

To carry out this exercise, calculate the mean and standard deviation of the distribution of *Scores*. To calculate the mean and standard deviation:

    Stat>Basic Statistics>Descriptive Statistics
      Variables: Scores   [double-click to transfer]
      Graphs: (none)
        OK
      OK

Next the distribution of *z* scores will be calculated from *Scores* using the Standardize command, which be placed in a new variable named *Zscores*. Finally, frequency tables for both *Scores* and *Zscores* will be generated using the Tally command. To calculate the *z* scores and place them in a new variable:

    Calc>Standardize
      Input column(s): Scores   [double-click to transfer]
      Store results in: **Zscores**   [type in]
      Subtract mean and divide by standard deviation: Yes
      OK

To generate the frequency tables:

> *Stat>Tables>Tally*
>> Variables: Scores Zscores   [double-click to transfer]
>> Display: Counts, Percents, Cumulative counts, Cumulative percents
>> OK

The output from the above command displays the frequency (Count), cumulative frequency (CumCnt), relative frequency (Percent), and cumulative relative frequency (CumPct) distributions. It is the latter distribution (CumPct) that gives the percentile rank for each score value in the distribution. Recall, however, that the program does not interpolate within a score value when duplicate cases exist. Also, percentile points (score values corresponding to specific percentiles) can be determined from the output table only for the specific percentile ranks (CumPct) actually reported there.

To print the output from the Session window:

> *File>Print Window*
>> Print Range: All
>> OK

Exit Minitab:

> *File>Exit*

On exiting, Minitab will ask if you want to save changes to the worksheet. This is not advised, because the standard scores can be easily regenerated if needed.

Minitab Program Output

Descriptive Statistics

Variable	N	Mean	Median	Tr Mean	StDev	SE Mean
Scores	40	70.75	73.00	70.72	14.20	2.25

Variable	Min	Max	Q1	Q3
Scores	46.00	97.00	56.50	81.50

Summary Statistics for Discrete Variables

Scores	Count	CumCnt	Percent	CumPct	Zscores	Count	CumCnt	Percent	CumPct
46	1	1	2.50	2.50	-1.74279	1	1	2.50	2.50
48	1	2	2.50	5.00	-1.60196	1	2	2.50	5.00
49	2	4	5.00	10.00	-1.53154	2	4	5.00	10.00
51	1	5	2.50	12.50	-1.39071	1	5	2.50	12.50
55	4	9	10.00	22.50	-1.10905	4	9	10.00	22.50
56	1	10	2.50	25.00	-1.03863	1	10	2.50	25.00
58	1	11	2.50	27.50	-0.89780	1	11	2.50	27.50
59	1	12	2.50	30.00	-0.82738	1	12	2.50	30.00
60	1	13	2.50	32.50	-0.75697	1	13	2.50	32.50
61	1	14	2.50	35.00	-0.68655	1	14	2.50	35.00
65	1	15	2.50	37.50	-0.40489	1	15	2.50	37.50
70	1	16	2.50	40.00	-0.05281	1	16	2.50	40.00
71	1	17	2.50	42.50	0.01760	1	17	2.50	42.50
72	3	20	7.50	50.00	0.08802	3	20	7.50	50.00
74	1	21	2.50	52.50	0.22885	1	21	2.50	52.50
75	2	23	5.00	57.50	0.29927	2	23	5.00	57.50
76	1	24	2.50	60.00	0.36968	1	24	2.50	60.00
78	4	28	10.00	70.00	0.51051	4	28	10.00	70.00
79	1	29	2.50	72.50	0.58093	1	29	2.50	72.50
80	1	30	2.50	75.00	0.65134	1	30	2.50	75.00
82	1	31	2.50	77.50	0.79218	1	31	2.50	77.50
83	1	32	2.50	80.00	0.86259	1	32	2.50	80.00
84	1	33	2.50	82.50	0.93301	1	33	2.50	82.50
85	2	35	5.00	87.50	1.00342	2	35	5.00	87.50
88	1	36	2.50	90.00	1.21467	1	36	2.50	90.00
91	1	37	2.50	92.50	1.42592	1	37	2.50	92.50
92	1	38	2.50	95.00	1.49633	1	38	2.50	95.00
93	1	39	2.50	97.50	1.56675	1	39	2.50	97.50
97	1	40	2.50	100.00	1.84841	1	40	2.50	100.00
N=	40				N=	40			

SPSS
(To accompany Table 4-1.)

SPSS will calculate frequency tables for raw scores and for z scores, but it will not calculate specific percentile ranks or points on request. It also does not interpolate when duplication of scores exists near a desired percentile, so answers may differ slightly from the sample answers in the *Guide*.

Start the SPSS program.

The data in Table 4-1 are:

97 93 92 91 88 85 85 84 83 82 80 79 78 78 78 78 78 76 75 75 74 72 72 72 71 70 65 61 60 59 58 56 55 55 55 55 51 49 49 48 46

The Data Editor window will fill the screen. Enter the 40 data values in the first column and label it with the variable name *Scores*. To enter the new variable name, point to the column heading and double click. The data you entered should appear on the spreadsheet as follows, in rows 1-40:

> Scores
> 97
> 93
> 92
> ⋮
> etc.

The Descriptives procedure can be used to calculate standard scores:

> *Statistics>Summarize>Descriptives*
>> Variable(s): Scores   [highlight and transfer]
>> Save standardized values as variables: Yes
>> OK

The standard scores are calculated and placed in column 2 of the spreadsheet, in a new variable called *Zscores*. To see the spreadsheet again:

> *Window>SPSS Data Editor*

The last step is to generate the frequency tables of the two variables:

> *Statistics>Summarize>Frequencies*
>> Variable(s): Scores, Zscores   [highlight and transfer]
>> Format:
>>> Order by: Descending values
>>> Continue
>> OK

This procedure will result in the original order, high to low, being retained in the frequency tables. Note that only intervals with one or more counts are displayed in the frequency tables that were output. The interval around the raw score of 47, for instance, is not displayed in the table because it had no frequencies. To determine the percentile rank for a raw score of 49, look at the cumulative percent

("Cum.") for the score value of 49, which is 10.0, indicating that a score of 49 is at the 10th percentile rank. Percentile points can be determined only for those percentiles listed in the table under "Cum." Remember, the program does not interpolate.

To print the output generated by the above:

> *File>Print*
> > Print range: All visible output
> > OK

To save the *Scores* data that you entered and the *Zscores* data as well:

> *Window>SPSS Data Editor*
> *File>Save as*
> > (Enter device and file name for SPSS save file.)
> > Save

Exit SPSS:

> *File>Exit SPSS*

SPSS Program Output

Descriptives

Descriptive Statistics

	N	Minimum	Maximum	Mean	Std. Deviation
SCORES	40	46.00	97.00	70.7500	14.2014
Valid N (listwise)	40				

Frequencies

Statistics

	N	
	Valid	Missing
SCORES	40	0
Zscore(SCORES)	40	0

SCORES

		Frequency	Percent	Valid Percent	Cumulative Percent
Valid	97.00	1	2.5	2.5	2.5
	93.00	1	2.5	2.5	5.0
	92.00	1	2.5	2.5	7.5
	91.00	1	2.5	2.5	10.0
	88.00	1	2.5	2.5	12.5
	85.00	2	5.0	5.0	17.5
	84.00	1	2.5	2.5	20.0
	83.00	1	2.5	2.5	22.5
	82.00	1	2.5	2.5	25.0
	80.00	1	2.5	2.5	27.5
	79.00	1	2.5	2.5	30.0
	78.00	4	10.0	10.0	40.0
	76.00	1	2.5	2.5	42.5
	75.00	2	5.0	5.0	47.5
	74.00	1	2.5	2.5	50.0
	72.00	3	7.5	7.5	57.5
	71.00	1	2.5	2.5	60.0
	70.00	1	2.5	2.5	62.5
	65.00	1	2.5	2.5	65.0
	61.00	1	2.5	2.5	67.5
	60.00	1	2.5	2.5	70.0
	59.00	1	2.5	2.5	72.5
	58.00	1	2.5	2.5	75.0
	56.00	1	2.5	2.5	77.5
	55.00	4	10.0	10.0	87.5
	51.00	1	2.5	2.5	90.0
	49.00	2	5.0	5.0	95.0
	48.00	1	2.5	2.5	97.5
	46.00	1	2.5	2.5	100.0
	Total	40	100.0	100.0	
Total		40	100.0		

Zscore(SCORES)

		Frequency	Percent	Valid Percent	Cumulative Percent
Valid	1.84841	1	2.5	2.5	2.5
	1.56675	1	2.5	2.5	5.0
	1.49633	1	2.5	2.5	7.5
	1.42592	1	2.5	2.5	10.0
	1.21467	1	2.5	2.5	12.5
	1.00342	2	5.0	5.0	17.5
	.93301	1	2.5	2.5	20.0
	.86259	1	2.5	2.5	22.5
	.79218	1	2.5	2.5	25.0
	.65134	1	2.5	2.5	27.5
	.58093	1	2.5	2.5	30.0
	.51051	4	10.0	10.0	40.0
	.36968	1	2.5	2.5	42.5
	.29927	2	5.0	5.0	47.5
	.22885	1	2.5	2.5	50.0
	.08802	3	7.5	7.5	57.5
	.01760	1	2.5	2.5	60.0
	-.05281	1	2.5	2.5	62.5
	-.40489	1	2.5	2.5	65.0
	-.68655	1	2.5	2.5	67.5
	-.75697	1	2.5	2.5	70.0
	-.82738	1	2.5	2.5	72.5
	-.89780	1	2.5	2.5	75.0
	-1.03863	1	2.5	2.5	77.5
	-1.10905	4	10.0	10.0	87.5
	-1.39071	1	2.5	2.5	90.0
	-1.53154	2	5.0	5.0	95.0
	-1.60196	1	2.5	2.5	97.5
	-1.74279	1	2.5	2.5	100.0
	Total	40	100.0	100.0	
Total		40	100.0		

CHAPTER 5

REGRESSION

CONCEPT GOALS

Be sure that you thoroughly understand the following concepts and how to use them in statistical applications.

- ◆ Regression constants, slope, intercept
- ◆ Scatterplot
- ◆ Regression line
- ◆ Least squares criterion
- ◆ Standard error of estimate

GUIDE TO MAJOR CONCEPTS

Straight Lines

Regression procedures are designed to predict the value of one
variable from knowledge of another variable. Specifically, _____   regression
permits us to determine the equation that best describes a straight-
line, or _____ , relationship, even if the two variables in question   linear
are not perfectly related (that is, even if all the points do not fall
exactly on the straight line).

    Suppose we make two measurements on each of three people.
Call the two variables X and Y. The simplest linear relationship
between these variables occurs when Y always equals X, which can
be expressed by the equation _____ . In that case, no matter what $Y = X$
the specific values of X and Y, Y will always equal X. Below is a table
for three subjects. Given the X values already presented in the table,
determine the Y values according to this equation, plot these values
on the graph provided (Fig. 5-1), draw the line passing though these
points, and label the line with its equation.

$Y = X$

Subject	Y	X
a		0
b		1
c		3

0
1
3

    Now consider a second relationship. Suppose Y is always twice
the value of X. This verbal rule can be expressed by the equation
_____ . Using the same X values as above, fill in the appropriate Y $Y = 2X$
values in the table below using this new equation.

$Y = 2X$

Subject	Y	X
a		0
b		1
c		3

0
2
6

Plot the new line on the same graph as the first (Fig. 5-1), and label it
with its equation.

    These two lines differ in **slope**, which we will symbolize by b
and which is the vertical distance divided by the horizontal distance
between any two points on the line. Consider the line $Y = X$ and the
two points, (1, 1) and (3, 3), which are on the line. To travel on the
graph between these two points, you must go up _____ units and then   2
to the right _____ units. Since the definition of b, the _____ , is   2; slope

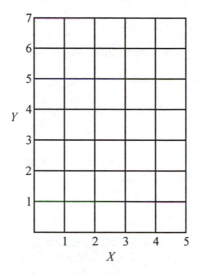

Fig. 5-1. Plot of three lines

the _____ distance between any two points on the line   vertical
divided by the _____ distance between them, we have in this   horizontal
case that

$$b = \underline{\hphantom{xxx}} = \underline{\hphantom{xx}}$$ $\dfrac{2}{2} = 1$

Now consider the line $Y=2X$ and the points $(1, 2)$ and $(3, 6)$
which lie on this line. To travel between these points on the line in
Figure 5-1, you must go vertically __ units and then horizontally   4
____ units. The slope for this line is   2

$$b = \underline{\hphantom{xxx}} = \underline{\hphantom{xx}}$$ $\dfrac{4}{2} = 2$

You will obtain the same value for the slope if you travel
between the two points in the opposite direction, from $(3, 6)$ to $(1, 2)$.
To do so, you must understand that positive values are at the top and
to the right in a graph, so when you travel upward or to the right you
are going in a _____ direction. However, when you travel   positive
downward or to the left, you are going in a _____ direction.   negative
Therefore, when you go from $(3, 6)$ to $(1, 2)$ you must travel
vertically (downward) ____ units and laterally (left) ___ units. Thus, $-4; -2$
the slope is

$$b = \underline{\hphantom{xxx}} = \underline{\hphantom{xx}}$$ $\dfrac{-4}{-2} = 2$

So we see that the slope of a line can be determined by using the
formula $b = \dfrac{y_2 - y_1}{x_2 - x_1}$ in which b is the slope and (x_1, y_1) and (x_2, y_2)

are any two points on the line. Using the formula $b =$ _____ the

$$\frac{y_2 - y_1}{x_2 - x_1}$$

slope for the line $Y = 2X$ given two other points (2, 4) and (4, 8) is

$$b = \rule{2cm}{0.4pt} = \rule{2cm}{0.4pt} = \rule{1cm}{0.4pt}$$

$$\frac{8 - 4}{4 - 2} = \frac{4}{2} = 2$$

Using the points in the reverse order, (4, 8) and (2, 4), the slope is

$$b = \rule{2cm}{0.4pt} = \rule{2cm}{0.4pt} = \rule{1cm}{0.4pt} = \rule{1cm}{0.4pt}$$

$$\frac{y_2 - y_1}{x_2 - x_1} = \frac{4 - 8}{2 - 4} = \frac{-4}{-2} = 2$$

Notice that the slope of the line $Y = 2X$ is 2, which is twice the slope of the line $Y = X$. Examine their equations:

$$Y = X$$

$$Y = 2X$$

The only difference between them is the 2 in the second equation. The number in front of the X is called its coefficient. When the equation of a line is in the form $Y = bX + a$, the slope of the line is always b, the coefficient of X. The value of b in the equation $Y = 2X$ is _____ , and in $Y = X$ it is _____ .

2; 1

Now consider a third possible relationship between Y and X. Suppose Y is always equal to twice the value of X plus 1. This verbal rule can be written mathematically as the equation _____ .

$Y = 2X + 1$

Below is a table containing the same X values as in the previous examples. Fill in the appropriate Y values using the above equation, plot these points on the same graph (Fig. 5-1), draw the line connecting these points, and label it with its equation.

$Y = 2X + 1$

Subject	Y	X
a		0
b		1
c		3

1

3

7

Look at the graph of this line and compare it with the graph of $Y = 2X$. Note that the line you just drew is always 1 unit higher than $Y = 2X$. It is customary to assess this difference along the vertical or Y-axis, where $X = 0$. The point where a line intersects the Y axis is called its **y-intercept**. By substituting $X = 0$ into the equation of the line, you can determine the value of the _____ .

y-intercept

Equation	Substituting $X = 0$	Value of Equation at y-Intercept
$Y = 2X$	$Y =$	$Y =$
$Y = 2X + 1$	$Y =$	$Y =$

2(0); 0

2(0) + 1; 1

Thus, the difference between the equations at $X = 0$ (i.e., at their
_____) is ____ unit. However, we could have seen this
fact by simply looking at the equations of the two lines; the value of
Y obtained from using a particular value of X in the second equation
will always be 1 more than the value of Y obtained for the same value
of X in the first equation.

y-intercepts; 1

$$Y = 2X$$
$$Y = 2X + 1$$

When the equation of a line is in the form $Y = bX + a$, the y-intercept
of the line is always a. For the two equations above, the y-intercepts,
symbolized by __ , are respectively __ and __ .

a; 0; 1

A straight line represents the relationship between two variables
X and Y, and the general equation for any straight line is given by
_____ , in which a is the _____ and b is the _____
of the line.

$Y = bX + a$; *y*-intercept; slope

Figure 5-2 presents graphs of three lines, A, B, and C. Determine
their equations:

A _____

B _____

C _____

$$Y = 3X + 2$$
$$Y = X - 1$$
$$Y = -2X + 10$$

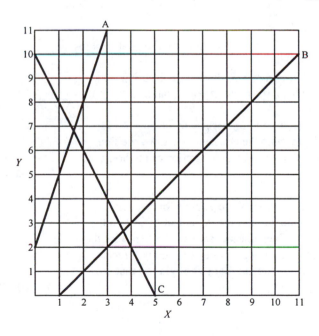

Fig. 5-2. Graph of three lines

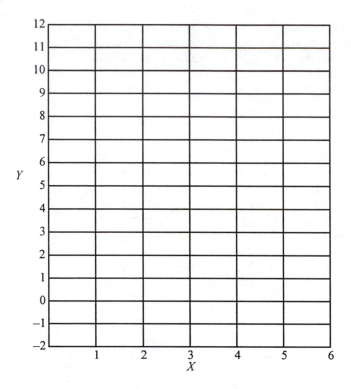

Fig. 5-3. Plot of three relationships

On the graph provided in Figure 5-3, draw lines describing the following relationships and label them with their equations:

1. A baby-sitter earns $5 an hour but must pay the agency $3 for each night of work. Determine the equation and draw the line representing the amount earned (Y) as a function of the hours worked (X). _____ $Y = 5X - 3$

2. Suppose that you rent a set of six wine goblets for a special dinner party. In addition to a rental fee, a deposit of $10 on the set of glasses is required. Each glass costs $2 if broken. Plot the relationship between the amount of money (Y) you will receive back from your deposit and the number (X) of glasses that are broken. _____ $Y = 10 - 2X$

Knowing the equation of a line helps not only to describe the relationship between two variables but to predict a specific Y value from knowledge of a specific X value. Determine the following values both by using the equations you determined immediately above and by drawing in Figure 5-3, at the appropriate X value, a vertical line that is perpendicular to the X-axis and that intersects the line of the relationship at the required Y value.

1. Using the relationship in number 1 above, how much will the baby-sitter earn in 4 hours? _____ $Y = 5(4) - 3 = \$17$

2. In number 2 above, how much will you get back from your $10 deposit if you or your guests break two glasses? _____ $Y = 10 - 2(2) = \$6$

Regression

In the graphs you have drawn in this chapter, all the points fall precisely on the line. The relationship between X and Y was errorless or _____ . When two variables from the behavioral sciences   perfect
are examined for the nature of their relationship, the observed data points (e.g., actual grades and SAT scores) rarely fall precisely along a straight line. A pictorial display of each pair of points in a relationship is called a **scatterplot**. The SAT scores and grade averages for 20 students are listed in Table 5-1. Plot these 20 points on the graph provided (Fig. 5-4) to form a _____ .   scatterplot

Table 5-1 SAT Scores and Grades for a Sample of 20 College Students

SAT	Freshman Grades	SAT	Freshman Grades
510	1.3	659	2.1
558	0.8	670	1.8
569	1.1	679	2.9
581	1.3	687	1.8
603	1.5	693	1.8
612	0.9	700	3.6
618	3.0	710	2.2
633	1.1	724	2.3
643	1.1	739	3.8
651	1.5	767	1.7

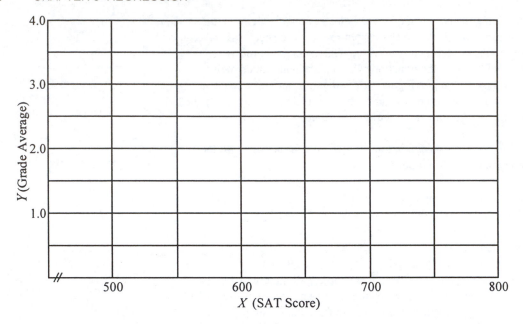

Fig. 5-4. Scatterplot of grade average as a function of SAT score.

Notice that the points show a generally increasing pattern but certainly do not fall precisely along a single straight line. One possible explanation of this is that, while there may indeed be a theoretically perfect linear relationship between scholastic assessment and college performance, there are many factors that produce variability in our measurements of these characteristics and thus cloud such a relationship. But if we had to fit a straight line to these points, what would be the best line we could draw? Statisticians define the best line as the one for which the sum of the squared vertical distances between the points and the line is as small as possible. This is called the criterion of _____ _____ .   least squares

To see how the **least squares criterion** works, it will be helpful to have a small sample of numbers such as those presented in Table 5-2. The line of relationship, called the _____ _____ ,   regression line
for this small set of data is $\hat{Y} = .9X = 1.5$, in which the symbol $\hat{Y}$ (read "Y predicted") is used because we can only estimate or predict Y imperfectly. Fill in the values that are missing from the first three columns of Table 5-2 by calculating $\hat{Y}$ for each X value (using the regression equation given above), calculating $\left(Y_i - \hat{Y}\right)$, and then squaring this difference. The sum of the column $\left(Y_i - \hat{Y}\right)$ should be zero (but rounding off numbers makes it .4 in this case). However, the sum of the column $\left(Y_i - \hat{Y}\right)^2$ will not usually be zero. In this case

it equals approximately _____ .

    Plot the regression line for these data in Figure 5-5. Next, place a point on the graph for each of the five points. Then draw a vertical dotted line between each Y_i point and the regression line, checking to see that the distance corresponds to the values for $Y_i - \hat{Y}$ that you calculated in Table 5-2. These vertical distances represent the error in the regression line, and the sum of their squared values should be as small as possible, according to the _____ _____ _____ .

6.7

least squares criterion

    Now, plot the line $\hat{Y} = X + 2$ on the same graph (Fig. 5-5), and draw with wavy lines the deviations between each point and this line. The new line looks as if it might be a reasonable summary of the imperfect linear relationship for these data. However, if the regression equation given initially is really the best line in the sense of least squares, then the sum of the squared differences, $\left(Y_i - \hat{Y}\right)^2$, between each point and the best regression line should be (larger/smaller) _____ than the corresponding sum computed for the alternative line, $Y = X + 2$. In short, over all five points, the dotted lines should be shorter than the wavy lines you have drawn in Figure 5-5. Calculate in Table 5-2 the values of $\hat{Y}_i$, $\left(Y_i - \hat{Y}\right)$, and $\sum\left(Y_i - \hat{Y}\right)^2$ for this alternative line. Is $\sum\left(Y_i - \hat{Y}\right)^2$ greater than 6.7?

smaller

_____ . Then the new line is not is good because the error is greater over the set of five points even if this alternate line fit three of the five points exactly. It is in this sense that the best fit line is defined as the one for which the sum of squared deviations, symbolized _____ , is a _____ .

Yes (it is 10)

$\sum\left(Y_i - \hat{Y}\right)^2$; minimum

Table 5-2 Illustration of the Least Squares Criterion

		The Regression Line, $\hat{Y} = .9X + 1.5$			An Alternative Line, $\hat{Y} = X + 2$		
X_i	Y_i	$\hat{Y}$	$\left(Y_i - \hat{Y}\right)$	$\left(Y_i - \hat{Y}\right)^2$	$\hat{Y}$	$\left(Y_i - \hat{Y}\right)$	$\left(Y_i - \hat{Y}\right)^2$
1	3						
4	3						
2	4						
7	9						
5	6						
			$\sum\left(Y_i - \hat{Y}\right) =$	$\sum\left(Y_i - \hat{Y}\right)^2 =$		$\sum\left(Y_i - \hat{Y}\right) =$	$\sum\left(Y_i - \hat{Y}\right)^2 =$

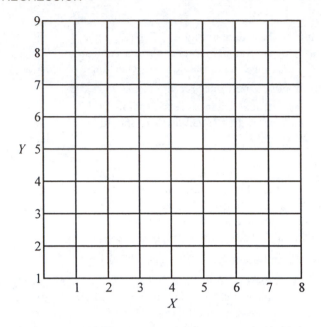

Fig. 5-5. The scatterplot and the two possible regression lines for the data in Table 5-2.

Standard Error of Estimate

We have seen that a regression line can be constructed even if the relationship between two variables is not perfect. The **standard error of estimate** is a measure of the **error** in predicting Y from a knowledge of X. In the discussion of the least squares criterion, it was said that the regression line is selected so that the sum of the squared deviations of the points from the line is a minimum. These squared deviations are also called the squared _____ of the points about the line, because the sum of squared deviations of points about the line is a measure of how *inaccurately* the regression line represents the relationship between two variables. A statistic that is roughly the average squared error per point in using the regression line is the _____ _____ _____ _____ ,

errors

standard error of estimate

symbolized by $s_{y \cdot x}$ and defined by

$$s_{y \cdot x} = \sqrt{\frac{\sum (Y_i - \hat{Y})^2}{N-2}}$$

To understand how this quantity works as a measure of _____,   error
consider the two sets of data given in Table 5-3. First plot the points
given in Table 5-3 on the graphs provided in Figure 5-6, which
already have regression lines drawn. (Note: these are not precisely
the true regression lines, but they will simplify computation.) Then,
draw with dotted lines the vertical distance between each point and
the line. This should give you a visual display of the relative error in
using the regression line as an index of the linear relationship in the
two sets of data. Now, in Table 5-3 calculate $(Y_i - \hat{Y})$, $(Y_i - \hat{Y})^2$,

$\sum(Y_i - \hat{Y})^2$, and $s_{y \cdot x}$ for each data set. For set A, $s_{y \cdot x} =$ _____ ;   1.29

for set B, $s_{y \cdot x} =$ _____ .   2.16

Table 5-3 Comparison of Standard Errors of Measurement

			Set A					Set B	
X_i	Y_i	$\hat{Y}$	$(Y_i - \hat{Y})$	$(Y_i - \hat{Y})^2$	X_i	Y_i	$\hat{Y}$	$(Y_i - \hat{Y})$	$(Y_i - \hat{Y})^2$
2	1	2			2	3	2		
4	5	4			4	2	4		
6	7	6			6	4	6		
8	7	8			8	10	8		
10	11	10			10	9	10		
$N =$			$\sum(Y_i - \hat{Y})^2 =$		$N =$			$\sum(Y_i - \hat{Y})^2 =$	

$$s_{y \cdot x} = \sqrt{\frac{\sum(Y_i - \hat{Y})^2}{N-2}} = \sqrt{} = \sqrt{}$$

$$s_{y \cdot x} = \sqrt{\frac{\sum(Y_i - \hat{Y})^2}{N-2}} = \sqrt{} = \sqrt{}$$

$s_{y \cdot x} =$ _____ $s_{y \cdot x} =$ _____

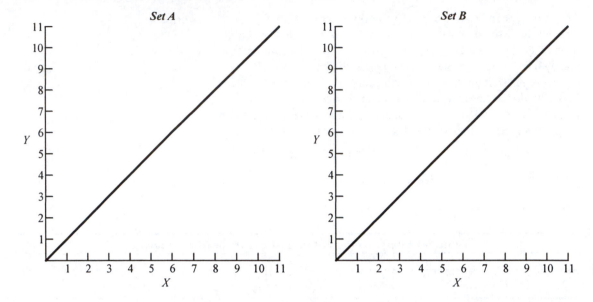

Fig. 5-6. Comparison of the scatterplots for two data sets.

Table 5-4 Guided Computational Example

Student	Test 1 X	Test 2 Y	X^2	Y^2	XY
a	3	1			
b	6	6			
c	4	5			
d	3	4			
e	8	8			
f	9	10			
g	2	3			
h	1	0			
i	4	5			
j	5	6			
k	2	8			
$N =$	$\sum X =$	$\sum Y =$	$\sum X^2 =$	$\sum Y^2 =$	$\sum XY =$
	$(\sum X)^2 =$	$(\sum Y)^2 =$			

Intermediate Quantities

$$(\mathbf{I_{XY}}) = N(\sum XY) - (\sum X)(\sum Y)$$
$$=$$
$$(\mathbf{I_{XY}}) =$$

$$(\mathbf{II_X}) = N\sum X^2 - (\sum X)^2$$
$$=$$
$$(\mathbf{II_X}) =$$

$$(\mathbf{III_Y}) = N\sum Y^2 - (\sum Y)^2$$
$$=$$
$$(\mathbf{III_Y}) =$$

Statistical Computations

$$\overline{X} = \sum X/N =$$

$$\overline{Y} = \sum Y/N =$$

$$s_x^2 = (\mathrm{II_X})/N(N-1) =$$

$$s_y^2 = (\mathrm{III_Y})/N(N-1) =$$

$$s_x = \sqrt{s_x^2} =$$

$$s_y = \sqrt{s_y^2} =$$

$$b = \frac{(\mathrm{I_{XY}})}{(\mathrm{II_X})} = \qquad\qquad =$$

$$a = \overline{Y} - b\overline{X} = \qquad\qquad =$$

$$\hat{Y} = bX + a = \qquad\qquad \text{For } X = 6, \ \hat{Y} =$$

$$s_{y\cdot x} = \sqrt{\left[\frac{1}{N(N-2)}\right]\left[(\mathrm{III_Y}) - \frac{(\mathrm{I_{XY}})^2}{(\mathrm{II_X})}\right]} =$$

Notice that the standard error of estimate is smaller for set A than for set B, indicating that the line approximates the location of the actual data points better for set ___ than set ___ . You should also be able to see from your scatterplot that the points cluster about the line more closely for set ___ than set ___ .

$A; B$

$A; B$

To summarize the computational routines necessary to calculate values for the concepts presented in this chapter, consider the following situation. A teacher gives two quizzes to a class. The scores of the 11 students are given in Table 5-4. Using the computational outline provided, determine the regression equation to predict scores on the second test from scores on the first, and calculate the standard error of estimate for this relationship. Use this Guided Computational format to solve future problems.

SELF-TEST

1. The equation for a regression line is $\hat{Y} = bX + a$ in which
   a. _____ is the predicted variable
   b. _____ is the predicting variable
   c. _____ is the y-intercept
   d. _____ is the slope

2. Find the slope and y-intercept of each of the following lines.

   a. $Y = X$ c. $2Y = 4X - 8$
   b. $Y = .5X + 2$ d. $X + 2Y = 6$

3. State what is known under the following circumstances:
   *a. The regression line always passes through the point (,).
   b. If $a = 0$, the regression line passes through the point with the coordinates (,).

c. If $b = 1$, what angle does the regression line have relative to the X-axis?

*d. If $\overline{Y} = 4$, $\overline{X} = 4$, and $a = 6$, is b positive or negative?

e. Describe the relationship if $s_{y \cdot x} = 0$.

f. If (4, 2) and (2, 4) are points on the line, is b positive or negative?

g. If a is negative and b is positive, the regression line crosses the X-axis to the (left/right) of the origin.

4. Define the following in words:

a. Y_i b. $\overline{Y}$ c. $\hat{Y}$

5. The least squares criterion selects the regression line such that what quantity is a minimum?

*6. Applying the regression equation is inappropriate when the relationship between X and Y is not _____ or when the value of X is outside the _____ of values upon which the regression equation was based.

*7. In a regression equation:

a. What units is a expressed in?

b. What units is b expressed in?

8. $s_{y \cdot x}$

*a. is expressed in which units, X or Y?

*b. can never be (greater/smaller) than s_y.

c. is based upon which deviations?

d. differs from the standard deviation of the Y scores (i.e., s_y) in which respects?

9. Indicate whether each of the following sets of facts could or could not possibly occur simultaneously. If the set is impossible, explain why.

a. $a = 12$, $b = 2$, $s_y = 1.5$, $s_{y \cdot x} = .5$

b. $a = -5$, $b = -1$, $s_{y \cdot x} = -2$

c. $s_y = 5$, $s_{y \cdot x} = 5$, $b = 2$

d. $Y = 20$ at $X = 12$, $s_{y \cdot x} = 2$, and (12, 28) is a typical score

*Questions preceded by an asterisk can be answered on the basis of the discussion in the text, but the discussion in this Study Guide does not answer them.

Table 5-5 Data for Exercises

Manual Dexterity	Spatial Skill	Job Success
24	6	2
41	10	9
35	7	6
28	4	4
35	11	5
48	4	10
40	8	4
28	7	5
38	9	9
36	6	4
29	10	1
33	9	5
25	8	5
25	8	6
37	6	9

EXERCISES

1. A personnel manager for an industrial firm gave job application tests for manual dexterity and visual spatial-relations perception to each of the last fifteen people hired for assembly positions. In addition, these employees have been rated on a scale from 1 to 10 (10 is the best score) for their success on the job after six months. These data are presented in Table 5-5. Calculate the means, variances, standard deviations, regression constants, and standard error of estimate: first, for predicting job success from manual

    dexterity and, second, for predicting job success from spatial-relations skill. Use the format of Table 5-4 as a computational guide.

2. June Ryan applies for a position as an assembler with the above company. She scores 38 on the dexterity test and 9 on the spatial test. Calculate the predicted job success rating from these two measurements. Which test is the more accurate predictor of job success and why?

ANSWERS

Table5-4. $N = 11$, $\sum X = 47$, $\sum Y = 56$, $\sum X^2 = 265$, $\sum Y^2 = 376$, $\sum XY = 297$, $(\sum X)^2 = 2209$, $(\sum Y)^2 = 3136$, $(\mathbf{I_{XY}}) = 635$, $(\mathbf{II_X}) = 706$, $(\mathbf{III_Y}) = 1000$, $\overline{X} = 4.27$, $\overline{Y} = 5.09$, $s_x^2 = 6.42$, $s_y^2 = 9.09$, $s_x = 2.53$, $s_y = 3.02$, $b = .90$, $a = 1.25$, $\hat{Y} = .90X + 1.25$, $\hat{Y}_{X=6} = 6.65$, $s_{y \cdot x} = 2.08$.

Self-Test. (1) $\hat{Y}$; X; a; b. **(2a)** 1; 0; **(2b)**.5, 2; **(2c)** 2; –4; **(2d)** –.5; 3. **(3a)** ($\overline{X}$, $\overline{Y}$); **(3b)** (0, 0); **(3c)** 45° angle; **(3d)** negative; **(3e)** all points fall on the regression line; **(3f)** negative; **(3g)** right. **(4a)** Y_i = the actual Y score for the ith subject; **(4b)** $\overline{Y}$ = mean of the observed Y scores; **(4c)** $\hat{Y}$ = predicted Y score at a given X, i.e., the regression line. **(5)** $\sum (Y_i - \hat{Y})^2$. **(6)** linear, range. **(7a)** Y units; **(7b)** b is unitless. **(8a)** Y units; **(8b)** greater; **(8c)** $\sum (Y_i - \hat{Y})^2$; **(8d)** deviations are taken about the regression line, not the mean, and the divisor is $N - 2$, not $N - 1$. **(9a)** Possible; **(9b)** impossible because $s_{y \cdot x}$ cannot be less than 0; **(9c)** impossible, because b must be 0 if $s_{y \cdot x} = s_y$; **(9d)** (12, 28) is not likely a "typical"

score since it would be 4 standard errors above the regression line.

Exercises. (1) Predicting job success from manual dexterity: $N = 15$, $\sum X = 502$, $\sum Y = 84$, $\sum X^2 = 17468$, $\sum Y^2 = 568$, $\sum XY = 2982$, $(\sum X)^2 = 252004$, $(\sum Y)^2 = 7056$; $(\mathbf{I_{XY}}) = 2562$, $(\mathbf{II_X}) = 10016$, $(\mathbf{III_Y}) = 1464$; $\overline{X} = 33.47$, $\overline{Y} = 5.6$, $s_x^2 = 47.70$, $s_y^2 = 6.97$, $s_x = 6.91$, $s_y = 2.64$, $b = .26$, $a = -2.96$ is exact, -3.10 if rounded numbers are used; $s_{y \cdot x} = 2.04$, $\hat{Y} = .26X - 2.96$. Predicting job success from spatial test: $N = 15$, $\sum X = 113$, $\sum Y = 84$, $\sum X^2 = 913$, $\sum Y^2 = 568$, $\sum XY = 624$, $(\sum X)^2 = 12769$, $(\sum Y)^2 = 7056$; $(\mathbf{I_{XY}}) = -132$, $(\mathbf{II_X}) = 926$, $(\mathbf{III_Y}) = 1464$; $\overline{X} = 7.53$, $\overline{Y} = 5.6$, $s_x^2 = 4.41$, $s_y^2 = 6.97$, $s_x = 2.10$, $s_y = 2.64$, $b = -.14$, $a = 6.67$ is exact, 6.65 if rounded numbers are used; $s_{y \cdot x} = 2.72$, $\hat{Y} = -.14X + 6.67$. **(2)** 6.92; 5.41; The manual dexterity test is the more accurate test, because its standard error is smaller.

STATISTICAL PACKAGES

StataQuest
(To accompany the guided computational example in Table 5-4.)

StataQuest will calculate the regression of a dependent variable onto a predictor variable as in the example shown in Table 5-4. The program will calculate the regression constants ("coef.") for the slope and y-intercept of the least squares regression line, and determine the standard error of estimate ("Root MSE"). It will also plot the least squares regression line. In a regression analysis, you must stipulate which variable is the "dependent" or Y variable, and which is the "independent" or X variable. In this example, the Y variable is *Test2*; it is predicted by the X variable, *Test1*.

Start the StataQuest program. Bring up the Stata Editor window and enter the scores for the predictor variable *Test1* in the first column, and the corresponding scores for the dependent variable *Test2* in the second column. Label the variable names. To enter the variable name, point to the column heading and double click. The data you entered should consist of eleven pairs of scores, which should appear on the spreadsheet as follows:

Test1	Test2
3	1
6	6
4	5
3	4
8	8
9	10
2	3
1	0
4	5
5	6
2	8

Save the data that you have entered:

File>Save as
    (Enter drive and file name for the save file.)
    OK

Open a log file to save or print the results of the analysis. Click on the *Log* button at the top left of the screen, enter a name and drive, and click OK.

Carry out the simple linear regression analysis:

Statistics>Simple regression
    Dependent variable: Test2   [click to transfer]
    Independent variable: Test1   [click to transfer]
    Show diagnostics menu: Yes
    OK
      Plot fitted model: Yes
      Run
      OK
   Cancel

On the output, the regression constants appear under the heading "Coef." along with other information that is not needed at this point. Specifically, the *y*-intercept is 1.247875, which is the "coefficient" of the "constant" term, and the slope is 0.8994334, which is the "coefficient" of *Test1*. The standard error of estimate ($s_{y \cdot x}$) is 2.0813, which is the last of the values listed in the upper right, labeled "Root MSE."

The regression equation, in this example, Test2 = 1.24788 + .899433*Test1, appears at the top of the plot.

  Print the regression output, both the log file and the graph, and exit StataQuest:

   File>Print Log
    OK

   File>Print Graph
    OK

   File>Exit

StataQuest Program Output

```
. regress Test2 Test1

  Source |      SS        df      MS              Number of obs =      11
---------+------------------------------          F( 1,     9) =   11.99
   Model | 51.9218388      1  51.9218388          Prob > F      = 0.0071
Residual | 38.9872521      9   4.3319169          R-squared     = 0.5711
---------+------------------------------          Adj R-squared = 0.5235
   Total | 90.9090909     10  9.09090909          Root MSE      = 2.0813


-------------------------------------------------------------------------
   Test2 |    Coef.   Std. Err.     t      P>|t|    [95% Conf. Interval]
---------+---------------------------------------------------------------
   Test1 |  .8994334   .259797    3.462    0.007    .3117319    1.487135
   _cons | 1.247875   1.275148    0.979    0.353   -1.63671     4.132461
-------------------------------------------------------------------------
```

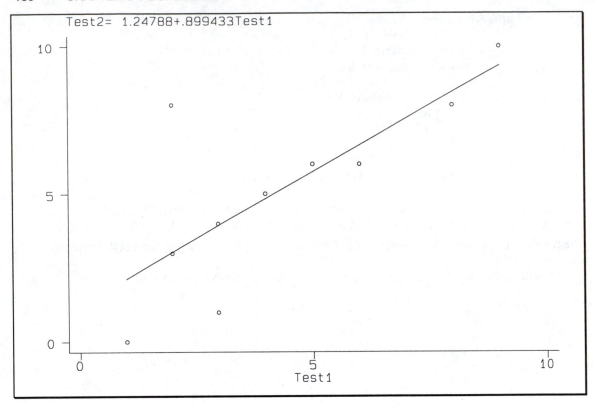

Test2= 1.24788+.899433Test1

Minitab
(To accompany the guided computational example in Table 5-4.)

Minitab will calculate the regression of a dependent variable onto a predictor variable as in the example shown in Table 5-4. The program will calculate the regression constants (called "coef") for the slope and y-intercept of the least squares regression line, and determine the standard error of estimate (called "S"). It will also plot the least squares regression line. The dependent variable is Y; the predictor variable X. In this example, the Y variable is *Test2*; it is predicted by the X variable, *Test1*.

Start the Minitab program. In the Data window enter the scores for the predictor variable *Test1* in the first column, and the corresponding scores for the dependent variable *Test2* in the second column. Label the variable names by typing them in at the top of the columns. The data you entered should consist of eleven pairs of scores, which should appear on the spreadsheet as follows:

Test1	Test2
3	1
6	6
4	5
3	4
8	8
9	10
2	3
1	0
4	5
5	6
2	8

Save the data that you have entered in a Minitab worksheet:

> *File>Save Worksheet As*
> > (Enter drive and file name for the save file.)
> > OK

Carry out the linear regression analysis and plot the least-squares estimate:

> *Stat>Regression>Fitted Line Plot*
> > Response (Y): Test2   [double-click to transfer]
> > Predictor (X): Test1   [double-click to transfer]
> > Type of Regression Model: Linear
> > OK

On the output, the regression equation, in this example, $Y = 1.25 + 0.899 * X$, appears at the top. The regression constants appear under the heading "Coef" along with other information that is not needed at this point. Specifically, the y-intercept is 1.248, which is the "coefficient" of the "constant" term, and the slope is 0.8994, which is the "coefficient" of X. The standard error of estimate ($s_{y \cdot x}$) is 2.081, which is labeled "S."

To print the regression output from the Session window, and to print the graph of the plot:

Window>Session
File>Print Window
       Print Range: All
       OK

Window>Fitted Line Plot
File>Print Window
       Print Range: All
       OK

Exit Minitab:

File>Exit

Minitab Program Output

```
Regression

The regression equation is
y = 1.25 + 0.899 x

Predictor        Coef        StDev           T           P
Constant        1.248        1.275        0.98        0.353
x               0.8994       0.2598       3.46        0.007

S = 2.081       R-Sq = 57.1%       R-Sq(adj) = 52.3%

Analysis of Variance

Source        DF          SS           MS           F           P
Regression     1        51.922       51.922       11.99       0.007
Error          9        38.987        4.332
Total         10        90.909
```

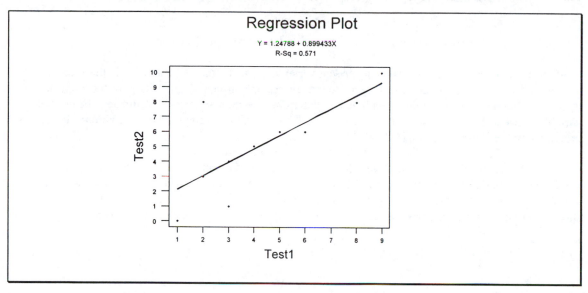

SPSS

(To accompany the guided computational example in Table 5-4.)

SPSS will calculate the regression of a dependent variable onto a predictor variable as in the example shown in Table 5-4. The program will calculate the regression constants (called "B") for the slope and y-intercept of the least squares regression line, and determine the standard error of estimate. The dependent variable is called Y; the predictor or independent variable, X. In this example, the Y variable is *Test 2*; it is predicted by the X variable, *Test1*.

Start the SPSS program. The Data Editor window should fill the screen.

Enter the scores for the predictor variable *Test1* in the first column, and the corresponding scores for the dependent variable *Test2* in the second column. Label the two columns with the variable names *Test1* and *Test2*. To enter a variable name, point to the column heading and double click. The data you entered should consist of eleven pairs of scores, which should appear on the spreadsheet as follows:

Test1	Test2
3	1
6	6
4	5
3	4
8	8
9	10
2	3
1	0
4	5
5	6
2	8

Carry out the simple linear regression analysis:

Statistics>Regression>Linear
        Dependent: Test2   [highlight and transfer]
        Independent: Test1   [highlight and transfer]
        Method: Enter
        OK

The output contains much more information than is needed at this point. In the output, you will get three tables labeled Model Summary, ANOVA, and Coefficients, respectively.

The regression constants appear in the Coefficients table under the heading "B." Specifically, the y-intercept is 1.248, which is the coefficient of the "Constant" term, and the slope is 0.899, which is the coefficient of *Test1*. The standard error of estimate ($s_{y \cdot x}$) is 2.0813, which is listed in the rightmost column of the Model Summary table.

Print the regression output:

> *File>Print*
> > Print range: All visible output
> > OK

Save the data that you entered:

> *Window>SPSS Data Editor*
> *File>Save as*
> > (Enter device and file name for SPSS save file.)
> > Save

Exit SPSS:

> *File>Exit SPSS*

SPSS Program Output

Regression

Model Summary[a,b]

Model	Variables Entered	Variables Removed	R	R Square	Adjusted R Square	Std. Error of the Estimate
1	TEST1[c,d]	.	.756	.571	.523	2.0813

a. Dependent Variable: TEST2
b. Method: Enter
c. Independent Variables: (Constant), TEST1
d. All requested variables entered.

ANOVA[a]

Model		Sum of Squares	df	Mean Square	F	Sig.
1	Regression	51.922	1	51.922	11.986	.007[b]
	Residual	38.987	9	4.332		
	Total	90.909	10			

a. Dependent Variable: TEST2
b. Independent Variables: (Constant), TEST1

Coefficients[a]

Model		Unstandardized Coefficients		Standardized Coefficients	t	Sig.
		B	Std. Error	Beta		
1	(Constant)	1.248	1.275		.979	.353
	TEST1	.899	.260	.756	3.462	.007

a. Dependent Variable: TEST2

CHAPTER 6

CORRELATION

CONCEPT GOALS

Be sure that you thoroughly understand the following concepts and how to use them in statistical applications.

- ♦ Correlation
- ♦ $\sum(Y_i - \overline{Y})^2$, $\sum(Y_i - \hat{Y})^2$, $\sum(\hat{Y} - \overline{Y})^2$
- ♦ The relation between r, s_y, and $s_{y \cdot x}$
- ♦ Effects on r of a restricted range of scores, of the use of extreme groups or combined groups, and of an extreme score

GUIDE TO MAJOR CONCEPTS

Correlation Coefficient

The square of the **correlation coefficient** is an index that reflects the proportion of variability in the Y variable that is associated with Y's linear relationship with variable X. This statistic was developed by Karl Pearson and is sometimes called the Pearson product-moment

_____ _____ .   correlation coefficient

    Consider Table 6-1. At the top is a scatterplot of five points that describe an approximately linear relationship between variables X and Y. The value of the points (i.e., Y_i), the mean of the Y's (i.e., $\overline{Y}$), and the value of the regression line $\hat{Y}_i$ at each X value are provided. The first column left vacant in the table at the bottom of Table 6-1 is for the simple deviations between each Y_i value and the mean of the Y's. This difference is symbolized by _____ . These deviations, $(Y_i - \overline{Y})$

when squared, summed, and divided by $N-1$, yield the _____   variance
of the Y_i, an expression of the total variability in the Y_i scores. Now, calculate $(Y_i - \overline{Y})$ for each score, and mark each deviation on the scatterplot with a straight line drawn between the point (Y_i) and the mean $(\overline{Y})$.

    Now consider only the point (9, 10) at the top right of the graph and in the first row of the table. The total distance between that point and the mean of the Y's is symbolized by _____. Numerically, $(Y_i - \overline{Y})$

it equals _____ = _____ . As you can see in the scatterplot, this $10 - 5 = 5$
total distance is composed of two parts, the distance from the mean to the regression line, symbolized by _____ , and the $(\hat{Y}_i - \overline{Y})$

distance from the regression line to the point, written _____ . $(Y_i - \hat{Y}_i)$

Fill in these symbol expressions in the appropriate parentheses at the right of the graph. Then determine the numerical value of these two distances for the point (9, 10) and write them in the first row of the table. Notice that $(Y_i - \overline{Y}) = (\hat{Y}_i - \overline{Y}) + (Y_i - \hat{Y}_i)$, which is numerically

verified, since ____ = ____ + _____ . On the graph, mark the $5 = 3 + 2$
distance between the mean and the regression line $(\hat{Y}_i - \overline{Y})$ with a wavy line and the distance between the regression line and the point $(Y_i - \hat{Y}_i)$ with a dotted line. Place these lines next to the continuous straight line you drew indicating the total distance, $(Y_i - \overline{Y})$.

Table 6-1 Numerical Illustration of the Correlation Coefficient

X_i	Y_i	$\overline{Y}$	$\hat{Y}_i$	Deviations			Squared Deviations		
				Total	Relationship	Error	Total	Relationship	Error
				$(Y_i - \overline{Y})$	$(\hat{Y}_i - \overline{Y})$	$(Y_i - \hat{Y}_i)$	$(Y_i - \overline{Y})^2$	$(\hat{Y}_i - \overline{Y})^2$	$(Y_i - \hat{Y}_i)^2$
9	10	5	8.0						
7	6	5	6.5						
5	1	5	5.0						
3	5	5	3.5						
1	3	5	2.0						
25	25			=_____	+_____		=_____	+_____	

Variance

$$\text{Variance of } Y = s_y^2$$

$$= \frac{\text{sum of total squared deviations of points about their mean}}{N-1}$$

$$= \underline{\qquad} = \underline{\qquad} = \underline{\qquad}$$

Correlation

$$\text{Correlation squared} = r^2 = \frac{\text{sum of squared deviations associated with } Y's \text{ relationship to } X}{\text{sum of total squared deviations of points about their mean}}$$

$$= \underline{\hspace{2cm}} = \underline{\hspace{2cm}} = \underline{\hspace{2cm}}$$

$$\text{Correlation coefficient} = r = \sqrt{\frac{\text{sum of squared deviations associated with } Y's \text{ relationship to } X}{\text{sum of total squared deviations of points about their mean}}}$$

$$= \sqrt{\underline{\hspace{2cm}}} = \sqrt{\underline{\hspace{2cm}}} = \underline{\hspace{2cm}}$$

Standard Error of Estimate

$$\text{Standard error of estimate} = s_{y \cdot x} = \sqrt{\frac{\text{sum of squared deviations of points associated with error}}{N-2}}$$

$$= \sqrt{\underline{\hspace{2cm}}} = \sqrt{\underline{\hspace{2cm}}} = \underline{\hspace{2cm}}$$

Now proceed *in the table* to the second point (7, 6). Determine algebraically the total deviation $(Y_i - \overline{Y})$ and then its two components $(\hat{Y}_i - Y)$ and $(Y_i - \hat{Y}_i)$, but watch the direction of the subtraction and the sign of your answer. Again, you should obtain the numerical equation $\underline{\hspace{1cm}} = \underline{\hspace{1cm}} + \underline{\hspace{1cm}}$. Draw these distances on the graph; notice that after you have drawn the wavy line for $(\hat{Y}_i - Y)$ you must "come back," or go down, from the line to the point $(Y_i - \hat{Y}_i)$, which is the same as adding the negative distance $-.5$. Fill in the columns of the table that call for the deviations of the remaining points, and draw the distances on the graph. Total the columns.

$$1 = 1.5 + (-.5)$$

The above process has illustrated that the total deviation between $\underline{\hspace{1cm}}$ and their $\underline{\hspace{1cm}}$ can be broken into two parts, the distance between the $\underline{\hspace{1cm}}$ and the $\underline{\hspace{1cm}}$ $\underline{\hspace{1cm}}$ plus the distance between the $\underline{\hspace{1cm}}$ $\underline{\hspace{1cm}}$ and the $\underline{\hspace{1cm}}$. This relationship can be expressed symbolically as

points; mean
mean; regression line
regression line
point

$$\underline{\hspace{1cm}} = \underline{\hspace{1cm}} + \underline{\hspace{1cm}}.$$

$$(Y_i - \overline{Y}) = (\hat{Y}_i - \overline{Y}) + (Y_i - \hat{Y}_i)$$

It is important to interpret these three parts conceptually. The differences between points and their mean $(Y_i - \overline{Y})$ represent the **total** deviation, as used in the common measure of variability, the

_____ . The distances between the regression line and the points $(Y_i - \hat{Y}_i)$ reflect the **error** in Y remaining after prediction of Y is made from the regression line. These distances contribute to the statistic called the _____ _____ ___ _____ . The distance between the mean and the regression line $(\hat{Y}_i - \overline{Y})$ represents the segment of the total deviation that is associated with Y's **relationship** to X. Thus, the total deviation between points and their mean can be divided into a part associated with Y's _____ to X and a part that remains as _____ in predicting Y from that relationship (i.e., from the regression line). A simple verbal equation expresses this relationship:

_____ = _____ + _____ .

However, as we have seen from our study of the variance, variability is usually expressed in *squared* deviations. One reason for this is that deviations that are not squared add up to zero over all the points. Did yours? Fill in the remaining three columns of Table 6-1, which require you to square the deviations that you have already filled in. Add the squared deviations in each column. Note that for squared deviations the equation "total = relationship + error" no longer holds for individual points; for the point (9, 10) for example, _____ ≠ _____ + _____ . However, for the *sums* of the columns (which are no longer zero), the equation "total = relationship + error" does hold: _____ = _____ + _____ .

At the bottom of the table you will find verbal expressions of the formulas for the variance, correlation, and standard error of estimate. Under each verbal statement, write the algebraic representation of the sum of the appropriate squared deviations and then the numerical value of the statistic for these data.

To check yourself: The total variability in Y_i can be expressed by the statistic called the _____ , which equals the sum of squared deviations for total, $\sum(Y_i - \overline{Y})^2 = $ _____ , divided by $N - 1 = $ _____ . So $s_y^2 = $ _____ / _____ = _____ .

The degree of relationship between X and Y can be expressed by the statistic called the _____ _____ . To obtain its value, determine the proportion of the total variability that is associated with Y's relationship to X by dividing the sum of squared deviations associated with the relationship, $\sum(\hat{Y}_i - \overline{Y})^2 = $ _____

by the total squared deviations, $\sum(Y_i - \overline{Y})^2 = $ _____ , which

proportion equals _____ / _____ = _____ . The square root of this value is the _____ _____ which equals _____ .

(right margin answers)

variance

standard error of estimate

relationship; error

total = relationship + error

$25 \neq 9 + 4$

$46 = 22.5 + 23.5$

variance

46

4; $46/4 = 11.5$

correlation coefficient

22.5

46

$22.5/46 = .49$
correlation coefficient
.70

The error in this relationship can be expressed by the statistic called the _____ _____ ____ _____ , symbolized by ____ . It equals the square root of the sum of squared deviations for error, $\sum (Y_i - \hat{Y})^2 =$____ , divided by $N - 2 =$_____ , which ratio equals _____ , and its square root, $s_{y \cdot x} =$_____ .

standard error of estimate

$s_{y \cdot x}$

23.5; 3

7.83; 2.80

In statistics, the formula that defines a concept is commonly not the most convenient formula to use for computation. The computational formula for the correlation coefficient is

$$r = \frac{N(\sum XY) - (\sum X)(\sum Y)}{\sqrt{\left[N \sum X^2 - (\sum X)^2 \right]\left[N \sum Y^2 - (\sum Y)^2 \right]}}$$

A guided computational example is presented in Table 6-2. Notice that you first calculate X^2, Y^2, and XY. After summing these quantities, determine the three intermediate quantities shown, which are the same intermediate quantities used in the previous chapter (Table 5-4) to obtain most of the statistics presented in Chapters 2-5. These quantities are to be substituted into the formula for r. Complete the calculation of the example.

Factors Affecting r

Since the correlation coefficient is an abstract, unitless index of the degree of relationship between two variables, it will be helpful to observe how its value changes under different conditions. Table 6-3 presents four examples. The scores in the X and Y distributions are the same in each of the four cases, but the X-Y pairings are different from case to case. Consequently, much of the computational labor has been eliminated, and only the numerator of r needs to be calculated (the denominator is always 200). Compute the value of r in each case, and plot the points on the graphs provided.

Case A illustrates a special and unlikely occurrence, a _____ relationship. In this case, all the points fall on the line, so Y_i and $\hat{Y}_i$ always have the same values. Therefore, $\sum (Y_i - \overline{Y})^2$ and $\sum (\hat{Y}_i - \overline{Y})^2$ are (equal/unequal) _____ , which means that the correlation coefficient, defined as

perfect

equal

$$r = \sqrt{\frac{\sum (\hat{Y} - \overline{Y})^2}{\sum (Y_i - \overline{Y})^2}}$$

will equal _____ .

1.00

Table 6-2 Guided Computational Example

Case	X	Y	X^2	Y^2	XY
a	7	4			
b	5	6			
c	6	4			
d	4	7			
e	2	5			
$N =$	$\sum X =$	$\sum Y =$	$\sum X^2 =$	$\sum Y^2 =$	$\sum XY =$
	$(\sum X)^2 =$	$(\sum Y)^2 =$			

Intermediate Quantities

$$(I_{XY}) = N(\sum XY) - (\sum X)(\sum Y) =$$

$$(II_X) = N\sum X^2 - (\sum X)^2 =$$

$$(III_Y) = N\sum Y^2 - (\sum Y)^2 =$$

$$r = \frac{N(\sum XY) - (\sum X)(\sum Y)}{\sqrt{\left[N\sum X^2 - (\sum X)^2\right]\left[N\sum Y^2 - (\sum Y)^2\right]}} = \frac{(I_{XY})}{\sqrt{(II_X)(III_Y)}} = \qquad =$$

A comparison of cases A and B illustrates the fact that the closer the points cluster about the line, the higher is the value of r (assuming that s_y is the same, which it is in these examples.) Draw the vertical lines connecting each point in B to the regression line to see this more closely. When squared, these deviations contribute to the statistic _____ . Therefore, under the special circumstances of this comparison, as the standard error of estimate becomes larger the correlation becomes _____ . Instead of $r = 1.00$ as in A, the $r =$ _____ in B.

$s_{y \cdot x}$

smaller
.80

Case C illustrates a situation in which there is almost no relationship between X and Y. Here, the regression line does not improve prediction very much, so the correlation is _____ . If the regression line did not aid prediction *at all*, the correlation coefficient would equal _____ .

$-.10$

0

Case D illustrates a _____ relationship; high Y values tend to be associated with (high/low) _____ X values. The $r =$ _____ .

negative
low; $-.80$

It should be observed that correlations of $-.80$ and $+.80$ represent (the same/a different) _____ degree of relationship; only the direction of the relationship is different.

the same

It is important to interpret the correlation coefficient in relation to the sample subjects that contributed data for its calculation. Table 6-4 presents some circumstances that can drastically alter the value of r. The age (X) and language proficiency score (Y) for 14 children are presented, and the subjects are divided into six subgroups, (a, b, c, d, e, and f) for the purpose of illustrating several facts. The N and the intermediate quantities ($\mathbf{I_{XY}}$), ($\mathbf{II_X}$), and ($\mathbf{III_Y}$) are given for the following four sets of subjects: (1) over all subgroups; (2) just over subgroups b, c, and d; (3) just over subgroups b, c, d, and f; and (4) just over subgroups a, b, d, e, and f. For each of these four sets of groups, draw the scatterplot and compute r now.

Case I, involving all 14 subjects, will serve as a reference group for the other situations. Since the correlation over the entire age range is _____ , you would conclude that there is a fairly high   .72
relationship between age and language performance. However, this relationship is limited to the ages studied. If the ages were 42 to 50 instead of 2 to 10, there might not be any relationship. Similarly, suppose that a teacher in a special school reads that the correlation between age and language ability is .72. As a result, the teacher decides to group the children, whose ages range from 4 to 8, into three age groups to make each class more homogeneous in language ability. However, consider just the data for four-, six-, and eight-year-olds (where we *restrict* the range to subgroups b, c, and d). Here in Section II of Table 6-4 you found the correlation to be only _____ ,   .31
quite a reduction from the value of ____ for the entire group. Thus the   .72
correlation coefficient is influenced by the range of values sampled; restricting the range of the variables often makes the correlation coefficient smaller.

Suppose that one mother begged the same teacher to let her 10-year-old child stay one more year in the special school. Suppose that this child is case f. What effect can one child have on a correlation? When you calculated r for the extreme-score example (subgroups b, c, d, and f) in Section III of Table 6-4, you found $r =$ _____ ,   .58
compared to _____ without that subject. Obviously, one   .31
subject with _____ scores can change a correlation quite   extreme
substantially.

Suppose that the original researcher who investigated the relationship between age and language has access only to 2- and 4-year-olds in a nursery school and 8- and 10-year-olds from two classes in a public school. The correlations you found in Section IV of Table 6-4 for such a situation (where we concentrate on the

Table 6-3 Some Examples of Correlation

A.

X	Y	XY
1	1	
3	3	
5	5	
7	7	
9	9	
25	25	

$$r = \frac{N(\sum XY)-(\sum X)(\sum Y)}{\sqrt{\left[N\sum X^2 -(\sum X)^2\right]\left[N\sum Y^2 -(\sum Y)^2\right]}}$$

$$r = \frac{5(\quad)-625}{200}$$

$$r =$$

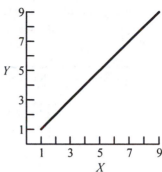

B.

X	Y	XY
1	1	
3	5	
5	3	
7	9	
9	7	
25	25	

$$r = \frac{N(\sum XY)-(\sum X)(\sum Y)}{\sqrt{\left[N\sum X^2 -(\sum X)^2\right]\left[N\sum Y^2 -(\sum Y)^2\right]}}$$

$$r = \frac{5(\quad)-625}{200}$$

$$r =$$

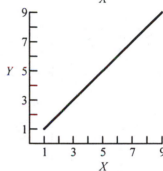

C.

X	Y	XY
1	3	
3	7	
5	9	
7	1	
9	5	
25	25	

$$r = \frac{N(\sum XY)-(\sum X)(\sum Y)}{\sqrt{\left[N\sum X^2 -(\sum X)^2\right]\left[N\sum Y^2 -(\sum Y)^2\right]}}$$

$$r = \frac{5(\quad)-625}{200}$$

$$r =$$

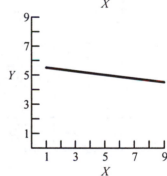

D.

X	Y	XY
1	7	
3	9	
5	3	
7	5	
9	1	
25	25	

$$r = \frac{N(\sum XY)-(\sum X)(\sum Y)}{\sqrt{\left[N\sum X^2 -(\sum X)^2\right]\left[N\sum Y^2 -(\sum Y)^2\right]}}$$

$$r = \frac{5(\quad)-625}{200}$$

$$r =$$

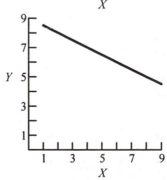

extreme groups a, b, d, e, and f) is _____ , compared to *r* for the   .82
total group of _____ , indicating that special sampling involving   .72
extreme groups can also influence *r*. It is clear that the value of the
correlation coefficient is highly dependent upon the distribution and
range of values included.

Table 6-4 Interpreting the Correlation Coefficient in Special Circumstances

Group	Age X	Language Score Y
a	2	1
	2	2
b	4	2
	4	3
	4	5
c	6	2
	6	4
	6	6
	6	7
d	8	3
	8	5
	8	6
e	10	6
f	10	9

Summary	N	(I_{XY})	(II_X)	(III_Y)
All groups	14	784	1232	969
b, c, and d	10	80	240	281
b, c, d and f	11	276	424	530
a, b, d, e, and f	10	560	880	536

I. All Groups (*a – f*)

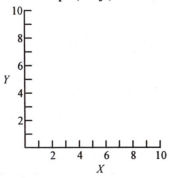

$$r = \frac{(I_{XY})}{\sqrt{(II_X)(III_Y)}}$$

$$r = \text{———————}$$

$$r = \text{———————}$$

Table 6-4 Interpreting the Correlation Coefficient in Special Circumstances (Continued)

II. Restricted Range (*b*, *c*, and *d*)

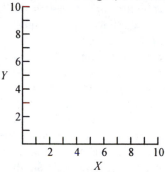

$$r = \frac{(\mathrm{I}_{XY})}{\sqrt{(\mathrm{II}_X)(\mathrm{III}_Y)}}$$

$$r = \underline{\hspace{2cm}}$$

$$r = \underline{\hspace{2cm}}$$

III. Extreme Score (*b*, *c*, *d*, and *f*)

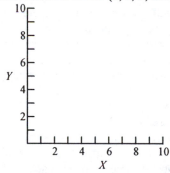

$$r = \frac{(\mathrm{I}_{XY})}{\sqrt{(\mathrm{II}_X)(\mathrm{III}_Y)}}$$

$$r = \underline{\hspace{2cm}}$$

$$r = \underline{\hspace{2cm}}$$

IV. Extreme Groups (*a*, *b*, *d*, *e*, and *f*)

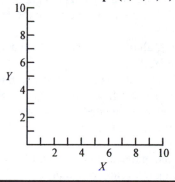

$$r = \frac{(\mathrm{I}_{XY})}{\sqrt{(\mathrm{II}_X)(\mathrm{III}_Y)}}$$

$$r = \underline{\hspace{2cm}}$$

$$r = \underline{\hspace{2cm}}$$

SELF-TEST

1. Which of the following measures reflects the variability in Y associated with differences in X? Which reflects the total variation in Y? Which reflects the variation in Y not associated with differences in X?

   a. $\sum(Y_i - \overline{Y})^2$

   b. $\sum(Y_i - \hat{Y})^2$

   c. $\sum(\hat{Y} - \overline{Y})^2$

*2. The correlation coefficient varies in size as a function of the relative sizes of s_y^2 and $s_{y \cdot x}^2$. Explain this relationship.

3. The value of r ranges between _____ and _____ .

4. When no linear relationship exists between X and Y, $r =$ _____ .

5. If $r = .60$, what proportion of the variability in Y is associated with the variability in X? What proportion is not associated with X?

*6. If $r = .70$ between age of grade-school children and broad jump distance in feet, what do you know about this relationship if jumping distance is measure in meters instead of feet?

*7. Under what circumstances is r equal to the slope, b, of the regression line of Y on X?

*8. What is the relationship, if any, between the correlation of X and Y and the correlation of Y and X?

9. Why might the correlation between scores on a test designed to predict success in college and subsequent college grades be lower for students at a very expensive private school than for students at State U?

10. The Smith Reading Diagnostic Test is reported to be a valid measure of reading skill because it correlated .75 with another measure of reading proficiency in a sample of 1000 children drawn from grades 1 to 6. Mrs. Jones, a reading specialist in the New York public schools, wants to use this test to select third-grade children who are having reading problems so that they can be given special training. What do you know about the validity of the Smith test for this purpose, that is, about the correlation between the Smith test and reading proficiency for third graders?

11. Suppose that within a sample of 100 male subjects an $r = .50$ is obtained between two variables. Indicate what is likely to happen to the size of r (will it increase, decrease, remain essentially unchanged, or change substantially in a direction that cannot be predicted?) in the following situations.

    a. N is increased to 150.
    b. The range of X values is restricted by eliminating some abnormally low scores.
    c. Two groups of subjects, one at each extreme of the X scale, are used as subjects.
    d. A single subject with extremely high positive X and Y scores is added to the sample.
    e. 10 subjects, all of whom scored $\overline{X}$ and $\overline{Y}$, are added.
    f. 50 female subjects, whose within-group r is also .50 but who have a different mean and variance than the males, are combined in the same sample.

12. Suppose that the number of hours a child spends watching television programs regarded as violent correlates .70 with the tendency of these children toward aggressive social behavior. Comment on the possibility of concluding that viewing violent programs causes aggressive social behaviors.

*13. What is r under the following conditions?

   a. $s_y = s_{y \cdot x} \neq 0$

   b. $s_y < s_{y \cdot x}$

   c. $s_{y \cdot x} = 0$ and $s_y \neq 0$

*Questions preceded by an asterisk can be answered on the basis of the discussion in the text, but the discussion in this Study Guide does not answer them.

EXERCISES

1.   It has often been suggested that artistic ability is incompatible with analytical reasoning. Below are data from a hypothetical study in which researchers examined the relationship between these two abilities by giving tests designed to measure analytical reasoning and artistic skills to each child in a sample of boys and girls. Calculate the correlation between these measures within each sex group according to the scheme of the guided computational example in Table 6-2.

Males		Females	
Analytical Reasoning Ability	Artistic Ability	Analytical Reasoning Ability	Artistic Ability
5	6	6	5
7	8	8	7
3	1	4	0
1	4	2	3
8	3	9	2
2	2	3	1
6	7	7	6

2.   Using the data in exercise 1, calculate the correlation for the two sexes combined into one sample. Explain any differences.

3.   Suppose that the scores for analytical reasoning and artistic ability were transformed into standard scores separately within each sex and variable. Would the correlations in problems 1 and 2 change? Explain.

ANSWERS

Table 6-2. $N = 5$, $\sum X = 24$, $\sum Y = 26$, $\sum X^2 = 130$, $\sum Y^2 = 142$, $\sum XY = 120$, $\left(\sum X\right)^2 = 576$, $\left(\sum Y\right)^2 = 676$; $(I_{XY}) = -24$, $(II_X) = 74$, $(III_Y) = 34$.

Table 6-3. $r = 1.00$, $.80$, $-.10$, $-.80$.

Self-Test. (1a) $\sum (Y_i - \overline{Y})^2$ reflects the total variability in the Y scores; **(1b)** $\sum (Y_i - \hat{Y})^2$ reflects the variability in Y remaining after predicting with the regression line (i.e., the error); **(1c)** $\sum (\hat{Y} - \overline{Y})^2$ reflects the variability in Y attributable to X. **(2)** $r^2 = 1 - \frac{s_{yx}^2}{s_y^2}$, so as $s_{y \cdot x}^2$

becomes small or s_y^2 becomes large, r will increase. **(3)** -1.00 and $+1.00$. **(4)** 0. **(5)** $.36$; $.64$. **(6)** $r = .70$. **(7)** If $s_x^2 = s_y^2$ as when X and Y are both standardized. **(8)** The two r's are the same value. **(9)** Since the expensive private school is presumable highly selective, its student range would be more narrow than State's. **(10)** Very little. Since reading is related to age, it is possible that the $r = .75$ for grades 1-6 is due to age and that the correlation within the third grade is quite different because of a restricted range of scores. **(11a)** Essentially no change; **(11b)** decrease; **(11c)** increase; **(11d)** increase; **(11e)** essentially no change; **(11f)** change but the direction depends on the particular values. **(12)** Not proved by this information, because correlation does not necessarily imply causality. **(13a)** .00; **(13b)** impossible; **(13c)** ±1.00.

Exercises. (1) For males: $(\mathbf{I}_{XY}) = 149$, $(\mathbf{II}_X) = 292$, $(\mathbf{III}_Y) = 292$; $r = .51$; for females: $(\mathbf{I}_{XY}) = 149$, $(\mathbf{II}_X) = 292$, $(\mathbf{III}_Y) = 292$; $r = .51$. **(2)** For sexes combined: $(\mathbf{I}_{XY}) = 547$, $(\mathbf{II}_X) = 1217$, $(\mathbf{III}_Y) = 1217$, $r = .45$; although the correlation within males is the same as within females, the correlation for the combined sample is smaller because of differences in the means for the two groups. **(3)** The r for each sex would not be affected, because when all scores are standardized the changes in the units and origins do not influence r. However, the r for the combined group would change, because some scores (i.e., males) would be transformed differently than other scores (i.e., females.)

STATISTICAL PACKAGES

StataQuest
(To accompany the guided computational example in Table 6-2.)

StataQuest will plot a scatterplot and calculate the correlation between two variables. The variables are paired in the same manner as in the simple regression analysis.

Start the StataQuest program. Bring up the Stata Editor window and enter the data from Table 6-2 for the two variables X and Y into the first two columns of the spreadsheet. Label the two columns with the variable names X and Y. To enter the variable name, point to the column heading and double click. The data you entered should appear on the spreadsheet as follows:

X	Y
7	4
5	6
6	4
4	7
2	5

Save the data that you entered:

> *File>Save as*
> (Enter drive and file name for the save file.)
> OK

Open a log file to save or print the results of the analysis. Click on the *Log* button at the top left of the screen, enter a name and drive, and click OK.

Generate the scatterplot for the joint distribution:

> *Graphs>Scatterplots>Plot Y vs. X*
> Y Axis variable: Y   [click to transfer]
> X Axis variable: X   [click to transfer]
> OK

Calculate the correlation coefficient between X and Y:

> *Statistics>Correlation>Pearson (regular)*
> Data variables: X Y   [click to transfer]
> OK

The output shows the correlation between X and Y to be -0.4785. The significance associated with the correlation (see Chapter 9) is $p = 0.4149$. Print both the log file and the graph, and exit StataQuest:

> *File>Print Log*
> OK

File>Print Graph
        OK

File>Exit

StataQuest Program Output

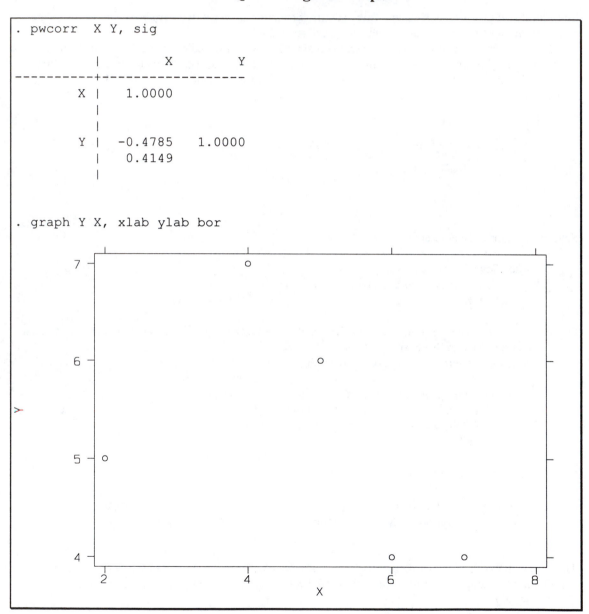

```
. pwcorr  X Y, sig

            |          X          Y
------------+------------------
        X |    1.0000
            |
            |
        Y |   -0.4785    1.0000
            |    0.4149
            |


. graph Y X, xlab ylab bor
```

Minitab
(To accompany the guided computational example in Table 6-2.)

Minitab will calculate the correlation between two variables and will produce a scatterplot. The variables consist of pairs of values in the same manner as for the simple regression analysis.

Start the Minitab program. In the Data window enter the five pairs of data from Table 6-2 for the two variables X and Y into the first two columns of the spreadsheet. Label the two columns with the variable names X and Y. To enter the variable names, type them in at the top of the columns. The data you entered should appear on the spreadsheet as follows:

X	Y
7	4
5	6
6	4
4	7
2	5

Save the data that you entered:

> *File>Save Worksheet As*
>> (Enter drive and file name for the save file.)
>> OK

Calculate the Pearson correlation between X and Y:

> *Stat>Basic Statistics>Correlation*
>> Variables: X Y   [double-click to transfer]
>> OK

The output shows the correlation between X and Y to be $-.478$. Minitab does not report the significance level associated with the correlation (see Chapter 9). Graph the scatterplot for the joint distribution:

> *Graph>Plot*
>> Graph variables:
>>> Graph 1,Y: Y   [double-click to transfer]
>>> Graph 1,X: X   [double-click to transfer]
>> OK

Print both the output from the Session window and the graph of the scatterplot:

> *Window>Session*
> *File>Print Window*
>> Print Range: All
>> OK

*Window>Plot 'X' * 'Y'*
File>Print Window
Print Range: All
OK

Exit Minitab:

File>Exit

Minitab Program Output

Correlations (Pearson)

```
Correlation of X and Y = -0.478
```

Plot 'Y' * 'X'

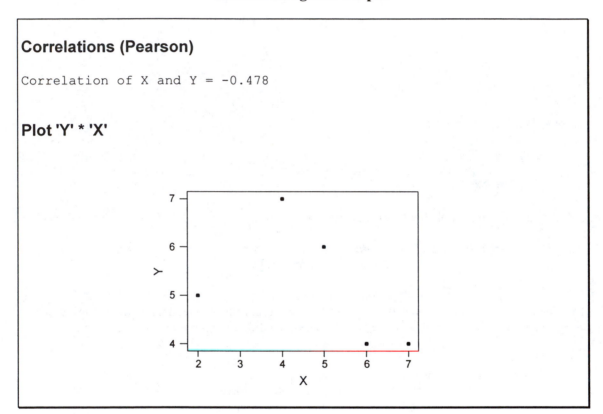

SPSS
(To accompany the guided computational example in Table 6-2.)

SPSS will calculate the scatterplot and correlation between two variables. The variables are paired in essentially in the same manner as for the simple regression analysis.

Start the SPSS program. The Data Editor window should fill the screen.

Enter the data for the two variables X and Y into the first two columns of the SPSS Data Editor spreadsheet. Label the two columns with the variable names X and Y. To define a variable name for a column, double click on the column heading. Then type in the variable name. The data you entered should appear on the spreadsheet as follows:

X	Y
7	4
5	6
6	4
4	7
2	5

Generate the scatterplot for the joint distribution:

Graphs>Scatter
    Simple:
        Define
        Y Axis: Y   [highlight and transfer]
        X Axis: X   [highlight and transfer]
    OK

Calculate the correlation coefficient between X and Y:

Statistics>Correlate>Bivariate
    Variables: X and Y   [highlight and transfer]
    Correlation coefficients: Pearson
    Test of significance: Two-tailed
    OK

The output shows the correlation between X and Y to be $-.478$. The significance associated with the correlation (see Chapter 9) is $p = .415$. To print the output:

File>Print
    Print range: All visible output.
    OK

To save the data that you entered:

Window>SPSS Data Editor
File>Save as
    (Enter device and file name for SPSS save file.)
    Save

Exit SPSS:

File>Exit SPSS

SPSS Program Output

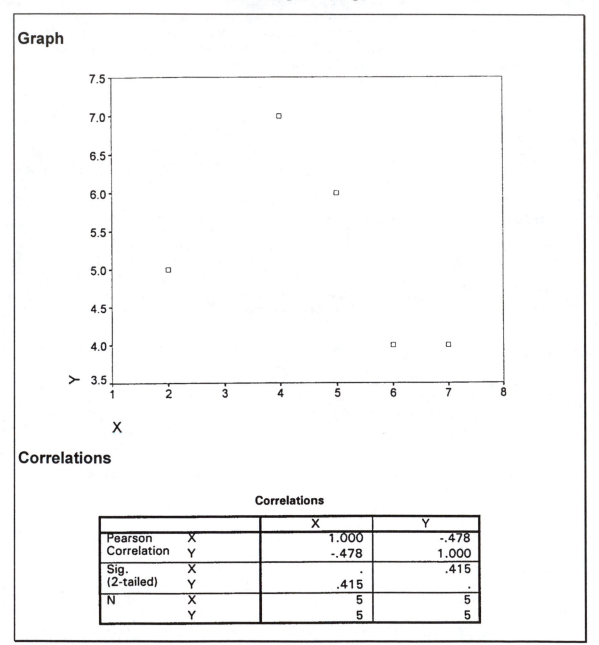

Graph

Correlations

Correlations

		X	Y
Pearson Correlation	X	1.000	-.478
	Y	-.478	1.000
Sig. (2-tailed)	X	.	.415
	Y	.415	.
N	X	5	5
	Y	5	5

PART 2

INFERENTIAL STATISTICS

CHAPTER 7

SAMPLING, SAMPLING DISTRIBUTIONS, AND PROBABILITY

CONCEPT GOALS

Be sure that you thoroughly understand the following concepts and how to use them in statistical applications.

- ◆ Population, parameter
- ◆ Sample, statistic
- ◆ Random and independent sampling
- ◆ Empirical and theoretical sampling distributions
- ◆ Standard error and its relationship to the standard deviation

GUIDE TO MAJOR CONCEPTS

Samples and Populations

When performing a research project, scientists usually observe only a **sample** of subjects or events selected from a larger **population**. After making the measurements, certain **statistics**, such as the mean and the variance, are often calculated on the sample and used to estimate their corresponding **parameters** in the population. Quantities and indices determined on a small _____ of subjects are known   sample
as _____ . The sample is only a subset of a larger   statistics
_____ whose characteristics are called _____ .   population; parameters
Research in the behavioral sciences often consists of calculating
certain _____ on a _____ of subjects for the purpose of   statistics; sample
estimating and inferring the values of corresponding
_____ in the _____ .   parameters; population

  Obviously, if statistics are to be good estimators of parameters, they must be based on a good sample of observations. Most frequently, a scientist attempts to obtain a **simple random sample**, one in which each subject or observation in the population has an **equal likelihood** of being selected and is picked **independently** of any other observation. Thus, in a _____ _____   simple random
sample, observations of subjects are selected _____ from   independently
one another and in such a way that each element of the population
has an _____ _____ of being chosen.   equal likelihood

Sampling Distributions

In addition to the population distribution and sample distribution of raw scores, one of the most important ideas in inferential statistics is that of a **sampling distribution**. A distribution of a statistic (the mean, for example) calculated on separate independent samples of size N from a given population is called a _____ _____ .   sampling distribution
Notice that it is not composed of unmodified direct observations; it is
a distribution of a _____ determined on numerous samples of such   statistic
observations. While the sample and population distributions are
composed of raw scores, a _____ _____ is the   sampling distribution
distribution of a _____ determined on each of several   statistic
independent _____ of raw scores taken from the   samples
_____ of raw scores.   population

  As a concrete example of a sampling distribution, we will now create an **empirical sampling distribution**. That is, we will actually calculate the mean for each of several samples selected from a single population of values and form a distribution of those means. Since

this distribution of means will actually be observed, it is called an
_____ _____ distribution. In contrast,   empirical sampling
most sampling distributions used by statisticians are **theoretical**.
Such distributions, based upon mathematical concepts as opposed to
actual observations, are called _____ _____   theoretical sampling
_____ .   distributions

    At the bottom of Table 7-1 are some numbers in boxes. Cut or
carefully tear these out and separate them (or make your own set on
another piece of paper). Suppose that these are raw scores for the
whole population; they are listed in the left column of the table. The
population distribution has a mean of 5 and a standard deviation of
2.16. Notice that Greek letters μ (mu) and σ (sigma) are used to
designate these quantities, because they are population
_____ , not sample _____ . A graphic display of   parameters; statistics
this frequency distribution is also presented in the table. Now put
your 12 population values into a hat or some other opaque container,
mix, and draw out a sample of four numbers at random by picking a
number, replacing it in the container, picking a second number,
replacing it, etc. Write the four numbers in the space under
"Observed Sample Distributions of Raw Scores"; calculate the mean,
$\overline{X}$, for this sample and write it in the third column of Table 7- 1.
That has already been done for sample *a*, so repeat the process until
you have a total of 10 samples. When you are finished, add up the 10
sample $\overline{X}$'s and obtain their average value ($\overline{X}_{\overline{x}}$). The formulas in
Table 7-1 for the mean and standard deviation, formerly written in
terms of the raw scores, X_i , are now expressed in terms of $\overline{X}_j$,
because the means, $\overline{X}_j$, are the scores in an empirical sampling
distribution of the mean.

    Now, look down the column labeled "Sampling Distribution of
the Mean." The figures under " $\overline{X}$ " are the means of the 10 samples
you collected. Since the sample mean is a statistic, this column
represents a distribution of a statistic and is thus an empirical
_____ _____ . Specifically, it is the   sampling distribution
_____ _____ of the _____. In the lower   sampling distribution; mean
right corner of the Table 7-1, place squares on the frequency
distribution graph for each of the 10 means, locating them over their
value on the abscissa; you can see that a new distribution (a
distribution of means) has been created.

Table 7-1 Creating an Empirical Sampling Distribution of the Mean

Population Distribution of Raw Scores	Observed Sample Distributions of Raw Scores ($N = 4$)	Sampling Distribution of the Mean $\overline{X}$	$\overline{X}^2$
1	a. (6, 1, 5, 6)	a. 4.5	
2	b. (, , ,)	b.	
4	c. (, , ,)	c.	
4	d. (, , ,)	d.	
4	e. (, , ,)	e.	
5	f. (, , ,)	f.	
5	g. (, , ,)	g.	
6	h. (, , ,)	h.	
6	i. (, , ,)	i.	
6	j. (, , ,)	j.	

$$\sum \overline{X} = \qquad \sum \overline{X}^2 =$$

$$\left(\sum \overline{X} \right)^2 =$$

Population column continued: 8, 9, 60

$$\mu_x = 5$$
$$\sigma_x = 2.16$$

$$\overline{X}_{\overline{x}} = \frac{\sum \overline{X}_i}{N} = \frac{\sum \overline{X}}{10} =$$

$$\sigma_{\overline{x}} = \frac{\sigma_x}{\sqrt{N_{\text{sample}}}} = \frac{2.16}{\sqrt{4}} = 1.08 \quad s_{\overline{x}} = \sqrt{\frac{N \sum \overline{X}_j^2 - \left(\sum \overline{X}_j \right)^2}{N^2 (N-1)}} = \sqrt{\frac{10 \sum \overline{X}_j^2 - \left(\sum \overline{X}_j \right)^2}{90}} =$$

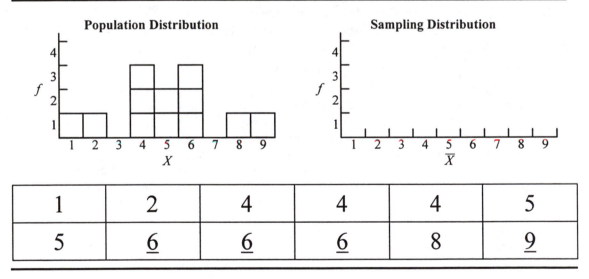

Population Distribution   Sampling Distribution

1	2	4	4	4	5
5	6	6	6	8	9

Now consider the task of estimating the mean of the population distribution of raw scores (supposing that it is too difficult to calculate directly). The population value is symbolized by μ_x and, in this case, we know that it equals _____ . We can estimate the mean of the population of raw scores by selecting a single sample of raw scores and calculating the sample _____ , symbolized by ____ . For the first sample, this value is _____ . For the second sample, it is _____ . But we can also estimate the mean of the population of raw scores, symbolized by ____ , with the average of the ten sample means which you have just calculated and which equals

_____ .

Here the concept of a **theoretical sampling distribution** is important. A theoretical sampling distribution is a distribution of a statistic over an uncountable number of samples. The theoretical sampling distribution of the mean is the distribution of the mean of an uncountable number of samples of a given size. The **mean** of the theoretical sampling distribution of the mean is symbolized by μ_x in which the μ indicates that the quantity is a _____ of a _____ _____ , and the subscript $\bar{x}$ denotes that it is the mean of a distribution of means (namely, of the _____ _____ of the _____). Since these two theoretical means are equal, a fact that can be expressed in symbols as _____ , there is no need to retain their subscripts, so it will be understood that μ without subscripts refers both to the mean of the _____ of _____ _____ and to the mean of the theoretical _____ _____ of the _____ . The empirical average of the ten sample means is represented by $\bar{X}_{\bar{x}}$, the mean of the means, and both $\bar{X}$ and $\bar{X}_{\bar{x}}$ are estimators of _____ .

5

mean; $\bar{X}$

4.5

$\left(\text{your } \bar{X}_b\right)$

μ_x

$\left(\text{your } \bar{X}_{\bar{x}}\right)$

mean
population distribution
sampling
distribution; mean
$\mu_x = \mu_{\bar{x}}$

population
raw scores
sampling distribution; mean

μ

Standard Error of the Mean

The sampling distribution of the mean has an average value; it also has a standard deviation. The standard deviation of the theoretical sampling distribution of the mean is symbolized by $\sigma_{\bar{x}}$, in which σ indicates a _____ _____ of a population distribution and the subscript $\bar{x}$ refers to the _____ _____ of the _____ . Statisticians call this quantity, symbolized by _____ , the **standard error of the mean**. Immediately below where you calculated the average of the 10 sample means, compute their standard deviation, $s_{\bar{x}}$, which will constitute an empirical estimate of the _____ _____ of the _____ .

standard deviation
sampling
distribution; mean
$\sigma_{\bar{x}}$

standard error; mean

There is a specific relationship between the *theoretical* standard error of the mean ($\sigma_{\bar{x}}$) and the theoretical standard deviation of the population of raw scores (σ_x):

$$\sigma_{\bar{x}} = \frac{\sigma_x}{\sqrt{N}}$$

This equation states that the theoretical standard error of the mean ($\sigma_{\bar{x}}$) for samples of size N equals the standard deviation of the population of raw scores divided by $\sqrt{N}$. If the standard deviation of the population of raw scores is 20, the standard error of the mean for samples of size 16 equals _____ . Thus, relative to the standard

$$\frac{20}{\sqrt{16}} = 5$$

deviation of raw scores, the standard error of the mean will always be (larger/smaller) _____ .

smaller

If the standard deviation of a single sample of scores (s_x) is an estimate of its corresponding population parameter, _____ , then s_x from a sample could be substituted for σ_x in this formula to give an estimate of the standard error of the mean:

σ_x

$$\sigma_{\bar{x}} = \frac{\sigma_x}{\sqrt{N}}$$

$$s_{\bar{x}} = \frac{}{\sqrt{N}}$$

s_x

Therefore, an estimate of $\sigma_{\bar{x}}$, written _____ , can be calculated on the basis of only one sample of cases, by dividing the _____ _____ of a single sample by the square root of _____ . For example, the standard deviation for the first sample of four cases in Table 8-1 equals 2.38. For this sample s_x = ___ and N = ___ so we estimate $\sigma_{\bar{x}}$ with

$s_{\bar{x}}$
standard
deviation; N

2.38; 4

$$s_{\bar{x}} = \underline{\hspace{2cm}} = \underline{\hspace{2cm}} = \underline{\hspace{2cm}}$$

$$\frac{s_x}{\sqrt{N}} = \frac{2.38}{\sqrt{4}} = 1.19$$

The standard error of a statistic is an index of **sampling error**. That is, the mean from one sample will probably not precisely equal the mean from another sample, as demonstrated by your calculations in Table 7-1. This variation from sample to sample in the value of a statistic is called _____ _____ . Since a standard deviation of a score represents its variability or the extent to which the scores deviate from one another, the standard deviation of the mean, called the _____ _____ __ _____ _____ , is an index of variability in the value of the mean from sample to sample, or

_____ _____ .

sampling error

standard error of the mean

sampling error

To summarize: A distribution of a statistic is called a _____

_____ . If that statistic is the mean, its distribution is called

the _____ _____ of the _____ . The mean of

the theoretical sampling distribution of the mean equals the mean of

the population of raw scores; this statement can be symbolized by

_____ . Therefore, both these quantities are represented simply by

_____ . The standard deviation of the theoretical sampling distribution

of the mean is symbolized by _____ and is called the _____

_____ of the _____ . Its relationship to the standard deviation

of the population of raw scores can be expressed by _____ .

Since the standard deviation of a single sample of scores, _____ ,

estimates _____ , the standard error of the mean may be estimated

from a single sample of N raw scores by the formula _____ .

	sampling
	distribution
	sampling distribution; mean
	$\mu_{\bar{x}} = \mu_x$
	μ
	$\sigma_{\bar{x}}$; standard
	error; mean
	$\sigma_{\bar{x}} = \dfrac{\sigma_x}{\sqrt{N}}$
	s_x
	σ_x
	$s_{\bar{x}} = \dfrac{s_x}{\sqrt{N}}$

Interpretation of Sampling Distributions

But what information does the standard error of the mean, $s_{\bar{x}}$,
convey? It is an expression of the accuracy of estimating the
population mean μ with a sample mean $\overline{X}$. Since the standard error
of the mean is the _____ _____ of the   standard deviation
sampling distribution of the mean, $s_{\bar{x}}$ estimates the extent to which
means of samples of size N vary one from another. The more similar
the value of a statistic from sample to sample, the (larger/smaller)
_____ its standard error. Therefore, the sample mean is   smaller
most accurate as an estimator of the population mean when its
_____ _____ is as (large/small) _____   standard error; small
as possible.

The sampling distribution of the mean and its standard error
possess two very important properties. The first is that regardless of
the form of the population distribution of raw scores, as the sample
size gets larger the sampling distribution of the mean becomes more
normal in form. For example, instead of drawing samples of size
$N = 4$ in Table 7-1, suppose that each of the 10 samples was of size
$N = 6$ or $N = 8$; then the larger sample size would make the
sampling distribution more _____ in form, even if the   normal
population distribution was highly skewed left or right. This fact is
important because it means that when the sample size is large the
percentiles of the normal distribution can be applied to the
_____ _____ of the _____ . (If the   sampling distribution; mean
population distribution is normal, then so is the sampling distribution
of the mean for *any* sample size.)

The second property is that the standard error of the mean becomes smaller as the size of the samples used to obtain the sampling distribution increases. If the 10 samples had each been of size $N = 6$ or 8 or 20, the size of $s_{\bar{x}}$ would have decreased as N (increased/decreased) _____ . Since the standard error reflects the extent to which the sample mean varies from one sample to another, this observation implies that means computed on small samples will tend to vary (more/less) _____ than means computed on larger samples. Therefore, X is a more accurate estimator of μ when it is determined on a (large/small) _____ sample.

increased

more

large

Applications

How can we use this information to make decisions about the results of our experiments? As an example, suppose that a psychologist and a physician know that a population of college students asked to remember 15 nouns can recall an average of 7 with a standard deviation of 2. Thus, the population parameters of this normal distribution are _____ and _____ . Now the scientists would like to know whether a certain drug has any influence on such memory performance. They first randomly select a person from the population, administer the drug, and record the number of correctly recalled nouns for this drugged subject. If the person's score falls at a very low percentile relative to the nondrugged population, then perhaps the drug has retarded memory performance. If the person scores at a high percentile, perhaps the drug has improved memory performance. Following scientific convention. the researcher decides that $P_{.025}$ and $P_{.975}$ will be cutoff limits; a score by the drugged individual equivalent to a percentile for non-drugged subjects between these boundaries will be regarded as usual nondrugged performance, while a percentile more extreme than these values will suggest that the drug has actually had an effect.

$\mu = 7$; $\sigma_x = 2$

Suppose the drugged person's actual score is 11. Since the population distribution is normal in form, this score can be translated into the standard normal deviate defined by:

$$z = \frac{X_i - \mu}{\sigma_x} = \underline{\hspace{2cm}} = \underline{\hspace{1cm}}$$

$\dfrac{11-7}{2} = 2.00$

Looking in Table A, Appendix 2 of your text, you can see that $z = $ _____ corresponds to a percentile rank of _____ , which is a very good performance. So the drug probably helps memory.

2.00; .9772

This result encourages the scientists to observe a group of 20 people performing the same task while under the influence of the drug. Suppose the group mean performance is 8.4. How can the scientists determine a percentile rank for this mean, since it is based on a group rather than on a single score? The theoretical sampling distribution of the mean is normally distributed; its mean is symbolized by ____ , and its standard deviation, symbolized by $\sigma_{\bar{x}}$,

μ

can be determined by the formula _____ . Thus, the

$$\sigma_{\bar{x}} = \frac{\sigma_x}{\sqrt{N}}$$

observed mean of 8.4 can be translated into a standard normal deviate with respect to the sampling distribution of the mean in essentially the same manner as above:

$$z = \frac{\bar{X} - \mu}{\sigma_{\bar{x}}} = \frac{\bar{X} - \mu}{\sigma_x/\sqrt{N}} = \underline{\hspace{2cm}} = \underline{\hspace{1cm}}$$

$$\frac{8.4 - 7}{2/\sqrt{20}} = 3.13$$

Looking in Table A, we see that a score of $z =$ _____ corresponds to a percentile rank of _____ , which is high enough for the scientists to conclude that the drug probably facilitates memory performance.

3.13
.9991

Suppose that an educator knows that in a given school system the mean reading score at the end of first grade is 72 with a standard deviation of 10, and 16 children who are instructed with a new reading program have a mean score of 78. If the new program is actually no better than the old one, at what percentile would this special sample fall in the sampling distribution for samples of size $N = 16$?

$$\mu = \underline{\hspace{2cm}} , \; \sigma_x = \underline{\hspace{1.5cm}}$$

72; 10

$$N = \underline{\hspace{0.8cm}} , \; \sigma_{\bar{x}} = \underline{\hspace{1.5cm}}$$

$$16; \; \frac{10}{\sqrt{16}} = 2.50$$

$$z = \frac{\bar{X} - \mu}{\sigma_{\bar{x}}} = \underline{\hspace{1.5cm}} = \underline{\hspace{1.5cm}}$$

$$\frac{78 - 72}{2.50} = 2.40$$

which has a percentile rank of _____ . Since this is beyond $P_{.975}$, it is likely that the new reading program is better than the old one.

.9918

Table 7-2 presents guided computational examples as a summary of the points made in this chapter.

Table 7-2 Guided Computational Examples

1. Given a population distribution with mean $\mu = 30$ and $\sigma_x = 6$, the mean $\mu_{\bar{x}}$ and standard deviation (standard error) $\sigma_{\bar{x}}$ of the theoretical sampling distribution of the mean for samples of size $N = 36$ are

$$\mu_{\bar{x}} = \mu_x = \mu = \underline{\hspace{2cm}}$$

$$\sigma_{\bar{x}} = \frac{\sigma_x}{\sqrt{N}} = \underline{\hspace{2cm}} = \underline{\hspace{2cm}}$$

2. If only a single sample of size $N = 36$ is available with mean $\overline{X} = 32$ and standard deviation $s_x = 8$, the parameters μ and $\sigma_{\bar{x}}$ may be estimated respectively by

$$\overline{X}_{\bar{x}} = \overline{X} = \underline{\hspace{2cm}}$$

$$s_{\bar{x}} = \frac{s_x}{\sqrt{N}} = \underline{\hspace{2cm}} = \underline{\hspace{2cm}}$$

3. The percentile rank of a *single score* $X_i = 43$ taken from the normal population in number 1 above can be determined from Table A of Appendix II with

$$z = \frac{X_i - \mu_x}{\sigma_x} = \underline{\hspace{2cm}} = \underline{\hspace{2cm}}$$

percentile rank = \underline{\hspace{2cm}}

So the probability of a randomly selected score being 43 or higher is $1.00 - .985 = .015$.

4. The percentile rank of a *mean* $\overline{X} = 27.25$ based on a sample of size $N = 16$ drawn from the population in number 1 above can be determined from Table A with

$$z = \frac{\overline{X} - \mu}{\sigma_{\bar{x}}} = \frac{\overline{X} - \mu}{\sigma_x / \sqrt{N}} = \underline{\hspace{2cm}} = \underline{\hspace{2cm}}$$

percentile rank = \underline{\hspace{2cm}}

So the probability of a sample of size $N = 16$ having a mean of 27.25 or more is $1.00 - .0336 = .9664$. This assumes that the population distribution of raw scores is normal in form or that N is large enough so that the sampling distribution of the mean is normal.

SELF-TEST

1. A quantity computed on a sample is a _____ , whereas that quantity determined on an entire population is called a _____ .

2. What is a simple random sample?

3. Explain the differences between the population distribution of raw scores, an observed sample distribution of raw scores, and a sampling distribution.

4. The standard deviation of a sampling distribution of means is called the _____ _____ .

5. Imagine a theoretical distribution of variances determined on all possible samples of size N drawn from a given population of scores. What would such a distribution be called, and what would be the special name given to the standard deviation of such a distribution?

6. What is the relationship between the mean of the theoretical sampling distribution of the mean and the mean of the population of raw scores? What is the relationship between the standard deviation of the theoretical sampling distribution of the mean and the standard deviation of the population of raw scores?

7. The variability of the theoretical sampling distribution of the mean will (always, often) be (smaller than, equal to, larger than) the variability of the population of raw scores.

8. Explain how it is possible to estimate the parameters of a sampling distribution of the mean by observing only one sample.

9. Under what conditions can one assume that the sampling distribution of the mean is normal in form?

*10. Under what conditions are the mean and variance of a theoretical sampling distribution independent of each other?

*11. What is an idealized experiment?

*12. What is the relationship between probability and theoretical relative frequency?

13. If a population of scores is normally distributed with $\mu = 50$ and $\sigma_x = 8$, determine the following probabilities.
   a. Sampling a score of 54 or lower, 54 or higher.
   b. Sampling a score of 24 or lower; 24 or higher.
   c. Obtaining a mean of 52 or higher with a sample of 16 observations; with a sample of 64 observations.
   d. Obtaining a mean of 46 or lower with a sample of size 36; with a sample of size 4.
   e. Obtaining a mean between 46 and 49 for a sample of size 16.

Questions preceded by an asterisk can be answered on the basis of the discussion in the text, but the discussion in this Study Guide does not answer them.

EXERCISES

1. Are the following procedures likely to produce a simple random sample of the stated population? Explain why or why not?
   a. There are six sections of Introductory Psychology being taught this semester. The classes are labeled one through six and a die toss is used to select which section is the sample of all introductory psychology students. The sample size is 30.
   b. Selecting the person listed in the upper right-hand corner of every fifth page of the telephone book as a sample of the telephone owners with listed numbers in the city.
   c. Selecting every third senior that answers an advertisement in the school paper as a sample of seniors.
   d. From a file of 150 questions, each one written on a card, selecting the first 20 questions after the cards were thoroughly shuffled as a sample of all the questions in the file.

2. Determine $\sigma_{\bar{x}}$ under the following conditions:
   a. $\sigma_x = 5$, $N = 16$
   b. $\sigma_x = 20$, $N = 36$
   c. $\sigma_x = 8$, $N = 64$
   d. $\sigma_x = 25$, $N = 100$

3. Suppose that a psychologist has developed a test to measure test anxiety in college students. The mean of a large population of students who have taken the test is 85 with $\sigma_x = 10$. The psychologist devises a test anxiety reduction technique that he teaches to a random selection of students. Tentatively assuming that knowing the technique has no effect on anxiety scores, what is the probability of a student who has received training scoring as low or lower than the following?
   a. 80   b. 92   c. 77   d. 70

4. Again assuming that the training in exercise 3 has no effect, what is the probability of obtaining a sample mean as low or lower than the following means, given the indicated sample sizes.
   a. $\bar{X} = 80$, $N = 9$
   b. $\bar{X} = 80$, $N = 16$
   c. $\bar{X} = 92$, $N = 25$
   d. $\bar{X} = 83$, $N = 100$

5. IQ is approximately normally distributed in the population with $\mu = 100$ and $\sigma_x = 16$. Suppose an educator obtains a truly random sample of three-year-old children and provides them with an intensive home education program administered by parents. After two years of the program, the researcher gives the children an IQ test. If the program has had no real effect on IQ, what is the probability of a group scoring as high as the following or higher?
   a. $\bar{X} = 108$, $N = 25$
   b. $\bar{X} = 104$, $N = 16$
   c. $\bar{X} = 104$, $N = 100$

ANSWERS

Table 7-2. (1) $\mu = 30$; $\sigma_{\bar{x}} = 1.00$. **(2)** $\bar{X} = 32$; $s_{\bar{x}} = 1.33$. **(3)** $z = 2.17$; 98.5th percentile. **(4)** $z = -1.83$; 3.36th percentile.

Self-Test. (1) Statistic; parameter. **(2)** Each observation in the population has an equal likelihood of being included in the sample, and the observations are selected independently of one another. **(3)** A population distribution of raw scores is a distribution of raw scores in a specified population; an observed sample distribution of raw scores is a distribution of raw scores from a sample from that population; a sampling distribution is a distribution of a statistic based upon repeated samples of size N from that population. **(4)** Standard error of the mean. **(5)** Sampling distribution of the variance, standard error of the variance. **(6)** $\mu_{\bar{x}} = \mu_x = \mu$, $\sigma_{\bar{x}} = \sigma_x / \sqrt{N}$. **(7)** Always; smaller (except of course when $N = 1$). **(8)** $\bar{X}$ estimates μ and $s_{\bar{x}} = s_x / \sqrt{N}$ (based upon one sample of size N) estimates $\sigma_{\bar{x}}$. **(9)** If the population of raw scores is normal or if the size of the sample is sufficiently large. **(10)** If the population distribution of raw scores is normal (or symmetrical) and the observations are independently and randomly sampled. **(11)** An idealized experiment is a hypothetical experiment composed of an unlimited number of identical repetitions of an event in which all natural factors contributing to the outcome of such events have an appropriate chance of influencing the result. **(12)** Theoretical relative frequency can be interpreted as probability. **(13a)** .6915, .3085; **(13b)** .0006, .9994; **(13c)** .1587, .0228; **(13d)** .0013,.1587; **(13e)** .2857.

Exercises. (1a) No, students were not independently selected, once one student was selected (the section), the entire sample was selected; **(1b)** yes; **(1c)** no, only seniors who responded to the advertisement were eligible for the sample; **(1d)** yes. **(2a)** 1.25; **(2b)** 3.33; **(2c)** 1.00; **(2d)** 2.50. **(3a)** .3085; **(3b)** .7580; **(3c)** .2119; **(3d)** .0668. **(4a)** .0668; **(4b)** .0228; **(4c)** .9998; **(4d)** .0228. **(5a)** .0062; **(5b)** .1587; **(5c)** .0062.

CHAPTER 8

INTRODUCTION TO HYPOTHESIS TESTING: TERMINOLOGY AND THEORY

CONCEPT GOALS

Be sure that you thoroughly understand the following concepts and how to use them in statistical applications.

- Assumptions for statistical tests:

  Random and independent sampling

  Normality of sampling distribution
- Null and alternative hypotheses
- Significance level
- Critical values and decision rules
- Theoretical relative frequency distributions z and t
- Sampling error
- Type I and type II decision errors
- Directional and nondirectional tests
- Degrees of freedom

GUIDE TO MAJOR CONCEPTS

Statistical Inference

Suppose a psychologist and a physician, knowing that a population of normal subjects recalls an average of 7 of 15 nouns ($\sigma_x = 2$) after an 80-minute delay, wondered whether administering a certain drug would influence memory performance in a group of 20 subjects. The researchers decided that if the mean of the drugged group fell between $P_{.025}$ and $P_{.975}$ in the non-drugged population, then the performance of the drugged group would be sufficiently typical of non-drugged people that it would not be interpreted as evidence that the drug had had an effect. However, if the performance of the drugged group was more extreme than these percentile ranks, then the researchers would conclude that the drug probably did have an effect. The mean for the group of drugged people was 8.4, and the standard normal deviate for this mean was

$$z = \frac{\overline{X} - \mu}{\sigma_x / \sqrt{N}} = \frac{8.4 - 7}{2 / \sqrt{20}} = 3.13$$

which has a percentile rank of .9991. Since this would be a very unusual and extreme level of performance for non-drugged individuals, the scientists concluded that the drug probably did influence performance. We now consider this example in greater detail.

Assumptions and Hypotheses

The statistical procedure used by scientists requires certain **assumptions** about the data. Statements about the population that are held to be true throughout the statistical analysis constitute its
_____ . The first assumption is that the subjects composing   assumptions
the drugged group were **randomly** and **independently** sampled from
the same population as the non-drugged subjects. For the statistical
procedures to give accurate probabilities, one must make the
_____ that the subjects in the drugged group were   assumption
_____ and _____ sampled from the same   randomly; independently
_____ as the non-drugged subjects.   population
     The second assumption is that the theoretical sampling
distribution of the means of non-drugged subjects is **normal** in form.
This assumption is necessary because the percentiles of the standard
normal distribution are not accurate unless the population

distribution of the statistic being evaluated (X) is _____ in   normal
form. Thus, the assumptions of this statistical technique are that the
subjects are _____ and _____ sampled from a   randomly; independently
common population and that the sampling distribution of the mean is
_____ in form.   normal

    Next, the logic of the process requires that the scientists state
some **hypotheses** about what might be true of this situation. Whereas
assumptions are held to be true throughout the statistical process and
are not tested, statements that represent a set of two or more
contradictory and often exhaustive possibilities, only one of which
can actually be the case, are known as _____ . For example,   hypotheses
one might hypothesize that the observed mean of 8.4 might be
reasonably typical of a sample of non-drugged individuals (i.e., the
drug has no demonstrable effect). Conversely, the observed
performance might be very atypical of the performance of non-
drugged people (i.e., the drug has an effect). Both of these
_____ cannot be true, so one is tentatively held to be true, and   hypotheses
this hypothesis is designated the **null hypothesis, H_0**. In this case,
the scientists temporarily suppose that the observed mean is typical
of non-drugged performance, that is, that the drug has no effect; this
supposition is formally stated as the _____ _____ .   null hypothesis
An **alternative hypothesis, H_1**, states that the drug does have an
effect. The null hypothesis is symbolized by ___ , while its alternative H_0
is designated ___ . In this case, H_1

    H_0: The observed mean is computed on a sample drawn from a
        population with $\mu = 7$ (i.e., the drug has no effect).

    H_1: The observed mean is computed on a sample drawn from a
        population with $\mu \neq 7$ (i.e., the drug has some effect).

    In summary, two _____ , which are usually mutually   hypotheses
exclusive and exhaustive, are stated. The one that is tentatively held
to be true is called the _____ _____ , symbolized   null hypothesis
by _____ . The other is designated the _____ hypothesis, H_0; alternative
represented by ___ . In the present case, the two hypotheses are: H_1

    H_0: _____   The observed mean is
computed on a sample drawn
from a population with $\mu = 7$.

    H_1: _____   The observed mean is
computed on a sample drawn
from a population with $\mu \neq 7$.

Significance Level

The question of whether a sample mean of 8.4 is typical or not
typical of the non-drugged population is decided by the **significance**

level (or **critical level**), symbolized by the Greek letter α (alpha).
The probability value that forms the boundary between rejecting and
not rejecting the null hypothesis is the _____ _____ ,   significance level
or _____ . That is, if it is probable that the sample mean of 8.4 comes α
from a population having $\mu = 7$ and $\sigma_x = 2$, then the null hypothesis
will not be rejected. Behavioral scientists usually choose a
significance level of .05 or one chance in 20 that such a result should
occur if the null hypothesis is true, although other values are possible
as well. Thus, if the probability is greater than .05 that H_0 is true, do
not reject H_0; otherwise, reject H_0. In this case, we say that the
_____ _____ is _____ ; in symbols, _____ .   significance; level; .05; $\alpha = .05$

Decision Rules

Such a probability value needs to be translated into terms that permit
a decision with respect to the null hypothesis. Such statements are
called **decision rules**, since they specify how high or low the sample
mean must be for H_0 to be rejected. Statements that designate the
statistical conditions necessary for rejecting the null hypothesis are
called _____ _____ . Decision rules are   decision rules
determined by obtaining the points in a theoretical relative frequency
distribution (e.g., the standard normal) corresponding to the
percentile ranks dictated by α. If the significance level is .05 and the
performance of drugged subjects could be better or worse than that of
non-drugged subjects, one needs the percentile points corresponding
to P_{025} and P_{975} of the standard normal distribution. These critical
values define a range which in the long run will contain 95% of the
means of samples from a population with $\mu = 7$ and $\sigma_x = 2$. The task
is to determine what values of z define the middle _____% of the   95
_____ _____ distribution. Thus, look at   standard normal
Table A in Appendix II of your text, not with a specific z value in
mind, but with a given proportion of area (i.e., probability) in mind.
To find P_{975}, scan the column giving proportions of area between the
mean and a specific z. We want to find the value _____ .   .4750
Alternatively, look down the column giving the proportions beyond a
given value of z, searching for the value _____ . The corresponding   .0250
z in either case is _____ . So the z value corresponding to P_{025} is   1.96;
_____ , and that corresponding to P_{975} is _____ . After transla-   −1.96; 1.96
ting the observed mean $\overline{X} = 8.4$ into a standard normal deviate—a
value which is called "z observed" or z_{obs}—we can state the statistical

conditions necessary for a decision on H_0, statements called _____
_____ . If _____ falls between _____ and _____ , we shall
not reject H_0; if it falls outside these values, the decision will be to
_____ _____ . We can state these rules using the
symbols z_{obs}, ± 1.96, $<$, $\leq$, and $\geq$, as follows:

If _____ , do not reject H_0.

If _____ or _____ ,
reject H_0.

decision

rules; z_{obs}; -1.96; 1.96

reject H_0

$-1.96 < z_{obs} < 1.96$

$z_{obs} \leq -1.96$; $z_{obs} \geq 1.96$

Computation and Decision

Given that the experiment has produced a group mean of 8.4, the
remaining task is one of **computation** and **decision**. First, the
observed mean must be translated into a _____ _____
_____ by the formula

standard normal
deviate

$$z_{obs} = \frac{\overline{X} - \mu}{\sigma_x / \sqrt{N}}$$

In this particular example, $\overline{X} =$ ___ , $\mu =$ ___ , $\sigma_x =$ ___ , and
$N =$ ___ , so we get the result

8.4; 7; 2
20

$$z_{obs} = \underline{\hspace{2cm}} = \underline{\hspace{2cm}}$$

$\dfrac{8.4 - 7}{2/\sqrt{20}} = 3.13$

Then z_{obs} must be compared with the decision rules: Does z_{obs} fall
within the limits of the first rule or the second? In this case,
$z_{obs} =$ ___ corresponds to the (first/second) _____ rule,
so H_0 is _____ .

3.13; second
rejected

What does it mean to *reject* H_0? It means deciding that the drug
very probably did have an effect on this type of memory
performance, because the statistical evidence is that a mean of 8.4 for
a sample of 20 subjects is very unlikely to occur by **sampling error**
alone. Sampling error must be considered because even if a group of
20 people was tested *without* receiving the drug, its mean likely
would not precisely equal the population mean, $\mu = 7$, nor would
its group mean likely equal the observed mean of a second or third
group of 20 non-drugged subjects. Samples of subjects do not all
have the same mean, and such variation in the mean from sample to
sample is attributed to _____ _____ . In the
previous chapter we noted that a measure of the amount of sampling
error is given by the _____ _____ of the mean.
In the present case, the standard error equals

sampling error

standard error

$$\sigma_{\overline{x}} = \sigma_x / \sqrt{N} = \underline{\hspace{2cm}} = \underline{\hspace{1.5cm}} .$$

$2/\sqrt{20} = .447$

Consequently, the rationale of the statistical test is to construct the sampling distribution of the mean for samples of size $N = 20$ and ask how likely it is that the observed mean could deviate that much from the population mean simply as a function of _____ _____. Since $\overline{X} = 8.4$ corresponds to a value of $z_{obs} = 3.13$ in the _____ _____ of the mean, a value which has a percentile rank of .9991, the observed mean is very unlikely to deviate from the population mean by _____ _____ alone. Thus, the null hypothesis is probably wrong, and the researchers decide to _____ _____ , concluding that the drug probably did have an effect.

    To summarize this basic process of statistical inference: first we must make the _____ that the subjects have been _____ and _____ sampled and that the sampling distribution of the mean is _____ in form. We then propose two contradictory _____ :

_____ : _____
    _____
    _____

_____ : _____
    _____
    _____

We tentatively adopt the _____ _____ , symbolized by _____ . Then we arbitrarily select a probability value, the _____ _____ , or ____ ; the customary value of α is ____ . This probability value marks the difference between a decision to _____ H_0 and to _____ _____ H_0. After stating this probability in terms of standard normal deviates, we can formalize the _____ _____ as follows:

    If _____ , do not reject H_0.

    If _____ or _____ , reject H_0.

After conducting the experiment, we translate the observed group mean into a _____ _____ _____ by the formula

$$z_{obs} = \underline{\hspace{2cm}}$$

If z_{obs} comes under the second decision rule, we decide to _____ ____ , because the observed mean is so different from the population mean of non-drugged subjects that such a discrepancy would not be likely on the basis of _____ _____ alone.

Answers:
sampling
error
sampling distribution

sampling error

reject H_0

assumptions
randomly; independently
normal
hypotheses
H_0: The observed mean is computed on a sample drawn from a population with $\mu = 7$.
H_1: The observed mean is computed on a sample drawn from a population with $\mu \neq 7$.
null hypothesis
H_0
significance level; α
.05
reject; not reject

decision rules
$-1.96 < z_{obs} < 1.96$
$z_{obs} \leq -1.96$; $z_{obs} \geq 1.96$

standard normal deviate

$$\frac{\overline{X} - \mu}{\sigma_x / \sqrt{N}}$$
reject
H_0

sampling error

Decision Errors

Statistical inference helps researchers make a decision in the face of partial evidence, but it does not guarantee that the decision is correct. **Decision errors** occur when the statistical process leads to a wrong decision either to reject or not to reject H_0. There are two kinds of

_____ _____ . A **type I** error occurs when the   decision errors
statistical process leads to rejection of H_0 when H_0 is actually
correct. In the example of the drugged subjects the statistical process
led to the decision to reject H_0, but that decision might be incorrect:
the drug may actually have had no effect. If this had been true, we
would have made a _____ _____ of the kind   decision error
called _____ . Given the logic of the statistical procedure, one   type I
should expect to make such mistakes occasionally. Even if the drug
had no effect, a small percentage of groups would still have mean
values more extreme than the limits set in the decision rules, as a
result of sampling error alone. That percentage would equal α. If
$\alpha = .05$, then when the null hypothesis is actually true, the statistical
procedures will lead us to make a mistake and incorrectly reject H_0
in _____% of the decisions. In short, given $\alpha = .05$, the probability   5
of a _____ error is _____ .   type I; .05
    The second kind of mistake that can occur is called a **type II**
error. When the null hypothesis is actually wrong but the statistical
process declares that H_0 should not be rejected, statisticians say that
a _____ error has occurred.   type II
    To review: there are two kinds of possible incorrect statistical
results, called _____ _____ . If the null   decision errors
hypothesis is actually true but the statistical process decided to reject
it, a _____ error has been committed. Conversely, if the null   type I
hypothesis is actually false but the statistical procedure decides not to
reject it, a _____ error has occurred.   type II

Directional Tests

In the example of drugged and non-drugged subjects, the researchers
could not reasonably predict in advance whether the drug might have
a facilitating or retarding effect on memory. Consequently, the
alternative hypothesis had to be phrased as
    H_1: The observed mean is computed on a sample drawn from a
        population with $\mu \neq 7$.
This means that H_0 would be rejected if the observation mean were
either very high or very low relative to the population mean. This is
called a **nondirectional** alternative and leads to a **two-tailed**
statistical test, because the direction of the possible result is not

specified and extreme values in either tail of the sampling distribution will result in rejecting H_0. Thus, if one cannot firmly predict whether the observed mean will be greater or smaller than the population mean, a _____ alternative and a _____ test are required.

nondirectional
two-tailed

However, suppose that on the basis of a popular theory or other experiments, the researcher could confidently predict that the drug would *not* retard memory but might help it. In this case, the alternative hypothesis would be

H_1: The observed mean is computed on a sample drawn from a population with $\mu > 7$.

This is a **directional** alternative, and it prompts a **one-tailed** test of H_0. In this event, an observed mean must be greater than P_{95} to reject H_0. P_{95} is selected instead of P_{975} because $\alpha = .05$ means that H_0 is to be rejected when the value of z_{obs} is in a range too extreme to be attained by more than 5% of the cases by sampling error alone. Since the drug is not likely to produce extremely low scores, drug-induced deviations are likely to be only positive. The entire critical region is thus placed in one tail of the distribution to keep the probability of a type I error at α, since the researchers know before the experiment that it is almost impossible to obtain an extreme value in the other tail. Therefore, P_{95} is the critical value for a _____ alternative (a _____ test) at $\alpha = $ _____ .

directional; one-tailed; .05

The *t* Distribution

In the example given in this and the previous chapter, the population standard deviation, σ_x, and the population standard error of the mean, σ_x were known values. This rarely happens in actual research practice. We have already seen that the sample $s_{\bar{x}}$ is an estimator of the population $\sigma_{\bar{x}}$. Thus, substituting $s_{\bar{x}}$ for $\sigma_{\bar{x}}$, the formula translating an observed mean into a standard score becomes

$$\overline{}$$

$$\frac{\overline{X} - \mu}{s_{\bar{x}}}$$

However, when $\sigma_{\bar{x}}$ is estimated by $s_{\bar{x}}$, the standard normal distribution is not the appropriate theoretical sampling distribution to use. A new distribution, the *t distribution*, is used to determine the probability that an observed mean could be obtained by sampling error alone. When the standard error of the mean is estimated with

_____ , the appropriate distribution and formula are

$s_{\bar{x}}$

$$\frac{}{} = \frac{}{}$$

$$t_{obs} = \frac{\overline{X} - \mu}{s_{\bar{x}}}$$

It happens that there is a different t distribution for every size of sample, or more accurately, for every number of **degrees of freedom** in the calculation of the statistic being examined. In this case, the number of degrees of freedom, df, is $N-1$. The critical values of t for directional and nondirectional tests for several different degrees of freedom are presented in Table B in Appendix 2 in your text. Look at that table now. Suppose a sample of 20 subjects was being considered; the particular t distribution to be used depends on the number of _____ _____ _____ , and in this case, $df =$_____$=$_____ . Not knowing in advance whether the drug would help or hinder memory, the scientists proposed a _____ alternative with $\alpha = .05$. The value in the table corresponding to these circumstances is _____ . Since the t distribution, like the z, is symmetrical about its mean of 0, the two critical values are _____ and _____ . The decision rules are

degrees of freedom
$N-1=19$

nondirectional
2.093

If _____ , do not reject H_0.

If _____ or _____ , reject H_0.

$-2.093; 2.093$
$-2.093 < t_{obs} < 2.093$
$t_{obs} \le -2.093; t_{obs} > 2.093$

Suppose this same experiment had yielded a sample mean of 9.1 and a sample standard deviation of 3, based upon $N = 25$. Since the standard error of the mean $s_{\bar{x}}$ equals the sample standard deviation divided by the square root of N ($s_{\bar{x}} = s_x/\sqrt{N}$) and the population mean is still $\mu = 7$, then

$$t_{obs} = \underline{\hspace{2cm}} = \underline{\hspace{2cm}} = \underline{\hspace{2cm}}$$

The decision would be to _____ _____ .

$$\frac{\bar{X} - \mu}{s_x/\sqrt{N}} = \frac{9.1 - 7}{3/\sqrt{25}} = 3.50$$
reject H_0

Similarly, suppose 16 subjects had been observed and a directional test conducted because other research indicated the drug should help (not hinder) memory. Also assume for this example that $\alpha = .01$. Therefore the critical value of t at $\alpha = .01$ would be _____ , and the decision rules would be

2.602

If _____ , do not reject H_0.

$t_{obs} < 2.602$

If _____ , reject H_0.

$t_{obs} \ge 2.602$

Finally, given a normal population with $\mu = 80$, what is the probability that a sample of 25 should have a mean as extreme as 71 (nondirectional) by sampling error alone if $s_x = 20$? Use $\alpha = .05$. Complete all the details of this problem in Table 8-1.

Table 8-1 Guided Computational Example

Hypotheses

H_0: _____

H_1: _____ (directional/non-directional)

Assumptions and Conditions

1. _____

2. _____

Significance Level

$\alpha =$ _____

Decision Rules

If _____ , do not reject H_0.

If _____ , reject H_0.

Computation

$t = \dfrac{\overline{X} - \mu}{s_x / \sqrt{N}} =$ _____ $=$ _____

$df = N - 1 =$ _____

Decision

(Reject/Do not reject) H_0.

SELF-TEST

1. What is wrong with the following statements?
   a. "The statistical test found the statistical assumptions to be invalid."
   b. "The statistical test rejected both hypotheses."
   *c. "The statistical test proved that the null hypothesis could not possibly be correct."

2. Why is it necessary to assume that the sampling distribution of the mean is normal?

3. Although direct knowledge of the shape of the sampling distribution of the mean is rare, its normality may be assumed under what two conditions?

4. Why is it necessary to hold tentatively that the null hypothesis is true?

5. What relationship does the significance level have to the probability of a type I error?

*6. Why do we never *accept* H_0?

7. What role does the sampling distribution play in the logic of hypothesis testing?

8. What is a critical value?

9. Define type I and type II errors.

*10. Define the *power* of a test.

11. When a directional alternative is appropriate, why is the entire critical region placed in one tail?

12. Under what circumstances is it necessary to use the t rather than z distribution?

*13. Define and give an example of degrees of freedom.

14. What is the critical value for a one-tailed t test of a score from a single group if $\alpha = .05$, $N = 17$, and $s_x = 3$?

Questions preceded by an asterisk can be answered on the basis of the discussion in the text, but the discussion in this Study Guide does not answer them.

EXERCISES

1. If the population average income for owners of Mamma Mia Pizza Shops is $40,000 with $\sigma_x = \$5500$, what is the probability that a sample of 121 owners who have had a special course on pizza store management should average $40,800 a year if the course really has no influence on income levels? Follow the general outline given in Table 8-1 in setting up and solving this problem. Use $\alpha = .05$, non-directional.

*2. The population distribution of annual incomes in exercise 1 is not likely to be normal in form. How did you handle this problem?

3. The population mean brain-weight at 90 days of age for a particular strain of inbred mice is 3.36 grams. A group of 25 of these mice is reared in a special environment designed to stimulate diverse kinds of learning. The 90-day mean brain-weight of these mice is 3.40 with $s_x = .05$. Following the outline in Table 8-1, test the hypothesis that the special environment has no effect on brain-weight. Use $\alpha = .05$.

4. Suppose it is known from a national survey that 18-year-olds rate their relationship with their parents to be 2.75 on a scale ranging from 1 (very poor) to 5 (very good). However, a sample of 16 adolescents who had recently been arrested for delinquent acts had an average rating of 1.97 with a standard deviation of 1.20. Previous research also discovered a similar difference. Test the hypothesis that there is no evidence that delinquents feel that their relationship with their parents is less good than the population of nondelinquents. Again follow the outline in Table 8-1 and use $\alpha = .05$.

ANSWERS

Table 8-1. Hypotheses: H_0: The observed mean is computed on a sample drawn from a population with $\mu = 80$. H_1: The observed mean is computed on a sample drawn from a population with $\mu \neq 80$ (nondirectional.) *Assumptions*: The subjects are randomly and independently sampled; the sampling distribution of the mean is normal. *Significance level*: $\alpha = .05$. *Decision rules*: Nondirectional test with $df = N - 1 = 24$. If $-2.064 < t_{obs} < 2.064$, do not reject H_0. If

$t_{obs} \leq -2.064$ or $t_{obs} \geq 2.064$, reject H_0. Computation: $t_{obs} = -2.25$; $df = 24$. *Decision*: Reject H_0.

Self-Test. (1a) The statistical procedures test the hypotheses, not the assumptions; **(1b)** Since the null and alternative hypotheses are mutually exclusive, they cannot both be false; **(1c)** The logic of the statistical test permits the rejection of the null hypothesis because a sample value was improbable, but even if a value is extremely unlikely, it is never shown to be impossible. **(2)** Normality of the sampling distribution of the mean is required to use the percentiles of the standard normal or t theoretical relative frequency distributions. **(3)** Normality is assumed if X is normally distributed in the population or if N is large, because the sampling distribution of the mean approaches normality as N becomes large. **(4)** Statistical logic requires that H_0 be held true and the statistical results examined to determine if the findings are consistent with H_0. **(5)** The significance level (α) is the probability of a type I error. **(6)** The statistical procedures tentatively assume H_0 to be true until there is evidence sufficient to contradict that hypothesis. Therefore, the evidence can only contradict H_0, because H_0 has already been *assumed* (not *proved*) true. **(7)** The sampling distribution and its standard error represent the extent to which statistics would be expected to vary from sample to sample on the basis of sampling error alone. **(8)** The critical value(s) define the boundary between rejecting and not rejecting H_0. **(9)** A type I error occurs when H_0 is actually true but the statistical decision is to reject it. A type II error occurs when H_0 is actually false but the statistical decision is not to reject it. **(10)** The power of a test is the probability of correctly rejecting H_0. **(11)** The entire critical region is placed in one tail to keep the probability of a type I error at α, because such an error is almost impossible to make in the other tail under these circumstances. **(12)** When $\sigma_{\bar{x}}$ is estimated with $s_{\bar{x}}$. **(13)** The number of degrees of freedom is the number of quantities that are free to vary when calculating a statistic. The df for $\overline{X}$ is N; for s_x it is $N-1$. **(14)** 1.746.

Exercises. (1) *Hypotheses*: H_0: The observed mean is computed on a sample from a population with $\mu = 40,000$. H_1: The observed mean is computed on a sample drawn from a population with $\mu \neq 40,000$. *Assumptions*: The subjects are randomly and independently sampled, and $\overline{X}$ is normally distributed. *Significance level*: .05. *Decision rules*: Given a non-directional test at .05, if $-1.96 < z_{obs} < 1.96$, do not reject H_0; if $z_{obs} \leq -1.96$ or $z_{obs} \geq 1.96$, reject H_0. *Computation*: $z_{obs} = 1.60$. *Decision*: Do not reject H_0. The evidence is not strong enough to conclude that the course is associated with a higher income level. **(2)** The sample size was large enough to assume that the sampling distribution of $\overline{X}$ was normal even if the population distribution of X was not. **(3)** Given a nondirectional test at .05 with $df = 24$, critical values are ± 2.064; $t_{obs} = 4.00$. Reject H_0. **(4)** Given a directional test at .05 with $df = 15$, the critical vale of t is 1.753. If $t_{obs} > -1.753$, do not reject H_0. If $t_{obs} \leq -1.753$, reject H_0. $t_{obs} = -2.60$. Reject H_0. (Note: Sometimes when testing directional hypotheses, the sequence of the means in the numerator is arranged to produce a positive difference. In this case, the signs and inequalities would be reversed accordingly in the critical values, decision rules, and α.)

CHAPTER 9

ELEMENTARY TECHNIQUES OF HYPOTHESIS TESTING

CONCEPT GOALS

Be sure that you thoroughly understand the following concepts and how to use them in statistical applications.

- ♦ t test of the difference between two independent group means
- ♦ t test of the difference between two correlated group means
- ♦ Test of the significance of r
- ♦ Test of the difference between two correlation coefficients

GUIDE TO MAJOR CONCEPTS

This chapter presents several techniques of statistical inference
commonly used in social science. Although the formulas differ from
one technique to another, it is important to realize that the basic logic
of the process is the same in each case. First, we assume for the entire
analysis that the data and population have certain characteristics.
These are the _____ of the analysis. Then we pose a set of   assumptions
mutually exclusive and usually exhaustive _____ about   hypotheses
the parameters of the population and tentatively hold one of the
hypotheses to be true. This is the _____ _____ . A   null hypotheses
probability level called the _____ _____ is   significance level
selected and used to specify a set of _____ _____ .   decision rules
Then we evaluate the observed data relative to an appropriate
_____ distribution, usually a known theoretical _____   sampling; relative
_____ distribution, such as the _____   frequency; standard
_____ or the _____ distribution. This z, t, or other statistic is   normal, t
compared to the _____ _____ to make the decision   decision rules
whether to _____ the _____ _____ .   reject; null hypothesis
This comparison reveals the probability that the observed results
would occur under the null hypothesis by _____ _____   sampling error
alone.

t Tests

Several approaches to be presented rely on the t distribution.
Although the specific terms of the formulas that translate the
statistics into t values differ, the general logic of the formulas is the
same. Recall from the previous chapter that

$$t = \frac{\overline{X} - \mu}{s_{\overline{x}}}$$

This expression is quite general: it could be written

$$t = \frac{(\text{a value}) - (\text{population mean of such values})}{(\text{estimate of the standard error of such values})}$$

That is, if the "value" is a sample statistic, then a sample statistic
minus the population mean of all such statistics divided by an
estimate of the standard error of this statistic is distributed as t. In the
previous chapter the sample statistic was the sample mean. The

population mean of sample means is symbolized by _____ , and an estimate of the standard error of such means is symbolized by _____ .

μ

$s_{\bar{x}}$

    But now suppose that the sample statistic is the difference between two sample means. Then the formula will read: the difference between two sample means minus the difference between those means in their _____ divided by an estimate of the _____ _____ of the difference between means is distributed as ____ . If the difference between two sample means is symbolized by $\overline{X}_1 - \overline{X}_2$, the difference between their population means by $\mu_1 - \mu_2$, and the estimated standard error of this difference by $s_{\bar{x}_1 - \bar{x}_2}$, then the formula will read

populations

standard error

t

$$t = \frac{(\quad\quad) - (\quad\quad)}{}$$

$$\frac{\left(\overline{X}_1 - \overline{X}_2\right) - \left(\mu_1 - \mu_2\right)}{s_{\bar{x}_1 - \bar{x}_2}}$$

The null hypothesis states that the two population parameters do not differ from each other, that is, that $\mu_1 = \mu_2$ and thus that $\mu_1 - \mu_2 =$ _____ . Therefore, under the null hypothesis the formula reduces to

0

$$t = \frac{\overline{X}_1 - \overline{X}_2}{s_{\bar{x}_1 - \bar{x}_2}}$$

and one needs only to determine the proper expression for $s_{\bar{x}_1 - \bar{x}_2}$.

Independent Groups

The problem discussed in the last chapter concerning the effect of a drug on memory performance would be better approached by having two separate samples that are treated exactly alike except that one group receives an injection of the drug and the other group receives an injection of a neutral saline solution. We could then directly compare the performance of the drugged group with that of the control group, which received the neutral saline solution. Statistically, that amounts to determining the probability that the two observed sample means differ from each other by _____ _____ alone. We shall consider this approach in detail.

sampling

error

    The basic question is whether the drug has an effect. We can translate this into two mutually exclusive and exhaustive statistical hypotheses, the **null hypothesis** and an **alternative hypothesis**. We want to decide whether the means of the populations of nondrugged and of drugged subjects are really identical. The proposition that they are identical, which we will tentatively assume to be true, is called the _____ _____ . It can be stated in symbols as

null hypothesis

$$\underline{\quad\quad} : \underline{\quad\quad\quad\quad}$$

$$H_0 : \mu_1 = \mu_2$$

On the other hand, these population means might not be equal. This is the _____ _____ , which we can write as _____ : _____

alternative hypothesis
$H_1 : \mu_1 \neq \mu_2$

The assumptions necessary to address these questions are (1) that the subjects in the two groups are **randomly** and **independently** sampled, (2) that the two groups are **independent** of each other, (3) that the population variances for the two groups are **homogeneous**, and (4) that the distribution of $\overline{X}_1 - \overline{X}_2$ is **normal** in form. If the subjects are to be typical of the population, they must be _____ and _____ sampled. Moreover, the two groups cannot be matched or related in any way; that is, the groups must be _____ of each other. If the groups are to be compared in the proposed manner, they must have the same variance in the population, a characteristic referred to as _____ of _____ . To use the percentiles of the t distribution, the distribution of $\overline{X}_1 - \overline{X}_2$ must be _____ , which will be the case if the two population distributions of scores are _____ or if the sample sizes are _____ .

randomly; independently

independent

homogeneity
variance
normal
normal
large

The next step is to determine the **degrees of freedom**, adopt a **significance level**, and state the **decision rules**. Suppose $N_1 = 15$ and $N_2 = 13$. The number of degrees of freedom for the statistical test to be made is $N_1 + N_2 - 2$ (see the text p. 239 for a discussion of this point). In this case the _____ _____ _____ equal _____ + _____ - _____ = _____ . Suppose we select .05 as the _____ _____ . We can now state the _____ _____ . Looking at the t distribution in Table B of Appendix 2 of the text, we see that a nondirectional test at the .05 level with $df = 26$ has critical values of _____ and _____ . If the observed value of t is between -2.056 and 2.056, the decision rule will be ___ _____ _____ _____ . In this case we would have insufficient evidence that the drug has had an effect. However, if the observed value of t is less than or equal to -2.056 or greater than or equal to 2.056, then we would decide to _____ _____ ; that is, we would conclude that the drug probably has an effect.

degrees of freedom
$15 + 13 - 2 = 26$
significance level
decision rules

-2.056; $+2.056$

do not reject H_0

reject H_0

A formal guided outline of this example is given in Table 9-1. In the appropriate places in that table, state the hypotheses, assumptions, and decision rules. Then complete the computational procedures outlined in the table.

Table 9-1 Guided Computational Example for a Test of the Difference between Two Independent Means

Hypotheses

H_0: _____

H_1: _____ (nondirectional/directional)

Assumptions and Conditions

1. The subjects are _____ and _____ sampled.
2. The groups are _____ .
3. The population variances are _____ .
4. The population distribution of _____ is _____ in form.

Decision Rules (from Table B of Appendix 2 in the text)
Given a significance level of _____ , a _____ test, and
$df = N_1 + N_2 - 2 =$ _____ = _____ :
If _____ , _____ .
If _____ , _____ .

Computation

	Drugged	Non-Drugged
	$N_1 = 15$	$N_2 = 13$
	$\overline{X}_1 = 9.1$	$\overline{X}_2 = 7.0$
	$s_1^2 = 9$	$s_2^2 = 4$

$$t_{obs} = \frac{\overline{X}_1 - \overline{X}_2}{\sqrt{\left[\dfrac{(N_1 - 1)s_1^2 + (N_2 - 1)s_2^2}{N_1 + N_2 - 2}\right]\left[\dfrac{1}{N_1} + \dfrac{1}{N_2}\right]}} =$$

Decision

_____ H_0 .

Since t_{obs} = _____ , which comes under the _____
decision rule, we _____ H_0. Thus, the data suggest that the
observed difference between the means of the drugged and non-
drugged samples was so large relative to what one would expect by
_____ _____ alone that the samples probably do
not come from populations having the same means. That is, the drug
is probably effective.

2.14; second
reject

sampling error

Correlated Groups

Consider another situation. An educator tests a sample of 12 nursery-school children to determine how well they grasp the concept of conservation of volume. This is tested by determining whether they realize that an amount of clay remains the same regardless of whether it is rolled into the shape of a long sausage or into a compact ball. The educator then institutes a program designed to give the children experiences that should improve their understanding of conservation of volume. After the training sessions, the same test that was initially given to the children is repeated. The children scored a mean of 6.0 on the first testing, but after the training the mean was 7.5. How likely is it that the mean score would have risen to 7.5 even without the training program, simply as a function of sampling error? That is, is the training program effective?

There is an important difference between the previous example and this one. Whereas before the measurements were made on two groups that were _____ , the pretraining and posttraining scores in this situation are assessed on the same subjects. Therefore, the groups of measurements are **correlated**. Whenever the measurements in the groups of scores are not independent but _____ , a slightly different statistical procedure for testing the null hypothesis is required.

The hypotheses are determined in the same way as for the previous case. The hypothesis to be tested, namely the _____ _____ , is that the population means for the first and second testings are identical. In symbols, _____ : _____ . The statement of the _____ _____ is slightly different in this particular problem. Whereas the previous example had a **nondirectional** alternative, this test has a **directional** alternative. Before, we wondered whether the drug had any effect, positive or negative. Thus the alternative hypothesis was _____ and was written ____ : _____ . In the present situation, no one would expect the children to perform less well on the second test than on the first one, so the alternative is said to be _____ . Given that μ_1 is the population mean on the first test and μ_2 the population mean on the second test, the alternative hypothesis can be written _____ : _____ .

The assumptions for this test are that the subjects have been _____ and _____ sampled and that the pairs of scores have been obtained from the same or related subjects. That is, the two sets of measurements are _____ . A third assumption is that the distribution of the differences between pairs of scores is _____ .

independent

correlated

null
hypothesis
$H_0 : \mu_1 - \mu_2$
alternative hypothesis

nondirectional; $H_1 : \mu_1 \neq \mu_2$

directional

$H_1 : \mu_2 > \mu_1$

randomly; independently

correlated

normal

If we adopt the .05 level of _____ with   significance

$N - 1 = 12 - 1 = 11$ _____ _____ _____ , we   degrees of freedom

determine the critical value by searching in the t table under a

_____ test at the _____ level with $df =$ ___ . Look up   directional; .05; 11

in Table B of your text this _____ _____ , which is _____ .   critical value; 1.796

You can now state two _____ _____ , being careful   decision rules

to remember that this is a _____ test. Therefore,   directional

_____ , _____ ,   If $t_{obs} < 1.796$; do not reject H_0

_____ , _____ .   If $t_{obs} \geq 1.796$; reject H_0

The computational procedure is somewhat different in the case
of correlated groups, and it is summarized in Table 9-2. Notice that
the posttraining and pretraining test scores are listed in that order,
since the second score is expected to be larger and the decision rule
for rejecting H_0 was for large positive t's. Then the scores are
subtracted, and the remainder is placed in the column D_i. Do this
now. Then square these values and place the results in the column
D_i^2. Now separately add the D_i and D_i^2 columns, compute $\left(\sum D_i\right)^2$,
and calculate the value of t_{obs} with the formula provided. The logic
of this procedure is that the differences between test scores, D_i,
should have a mean that approaches zero in the long run if there is no
difference in the populations from which the samples were drawn.
The above formula yields a t_{obs} which can be used to obtain the
probability that on the average the ___ deviate from _____ by D_i; zero

_____ _____ alone.   sampling error

The value of t_{obs} for these data is _____ , which is consistent   2.45

with the decision rule to _____ _____ . The teaching program   reject H_0

apparently improved performance of the children. Finish Table 9-2
now, and use it as a guide for solving future problems of this type.

Significance of *r*

Suppose a researcher takes a random sample of 42 women between
18 and 50 years of age and gives them a questionnaire to assess their
attitudes toward women's roles in society. Higher scores are
interpreted to indicate a more feminist attitude. The researcher
observes a correlation of $-.33$ between age and a feminist attitude;
that is, younger women tend to receive higher scores than older
women. The researcher wants to know the probability that such a
correlation is a chance result, and that the population correlation is
actually $\rho = .00$. That is, the value of $-.33$ observed for *r* may be

Table 9-2 Guided Computational Example for a Test of the Difference between Two Correlated Means

Hypotheses

H_0: _____

H_1: _____ (nondirectional/directional)

Assumptions and Conditions

1. The subjects have been _____ and _____ sampled.
2. The scores of the two groups are _____ .
3. The population distribution of D_i is _____ (i.e., X_1 and X_2 distributions are normal or N is large).

Decision Rules (from Table B of Appendix 2 in the text)

Given a significance level of _____ , a _____ test, and

$df = N - 1 =$ _____ $=$ _____ :

If _____ , _____ .

If _____ , _____ .

Data and Computation

Subject	X_2 Second Test	X_1 First Test	$D_i = (X_{2i} - X_{1i})$	D_i^2
a	6	5		
b	7	6		
c	3	4		
d	6	4		
e	2	2		
f	3	1		
g	4	2		
h	4	5		
i	7	5		
j	4	5		
k	9	7		
l	6	3		

$N =$ _____ $\sum D_i =$ _____ $\sum D_i^2 =$ _____

$(\sum D_i)^2 =$ _____

$$t_{obs} = \frac{\sum D_i}{\sqrt{\dfrac{N \sum D_i^2 - (\sum D_i)^2}{N - 1}}} = \underline{\hspace{2cm}} = \underline{\hspace{2cm}}$$

Decision

_____ H_0.

merely a consequence of _____ _____ .   sampling error

The hypotheses are that the correlation in the population is zero
versus that it is not zero. Specifically, the null hypothesis is
_____ , while the alternative hypothesis is _____ . One $H_0: \rho = 0$; $H_1: \rho \neq 0$
must assume that the subjects are _____ and _____   randomly; independently
sampled and that the distributions of both X (attitude score) and Y
(age) are _____ in form.   normal

The critical values of r for various df are listed in Table C in
Appendix 2 of the text. Turn to Table C now, and observe that it is
very similar in form to the t table. To use it, first decide whether the
alternative hypothesis is _____ or _____ and adopt a   directional; nondirectional
_____ _____ . In the present example, the alternative   significance level
hypothesis is that the population correlation is not .00, so a
_____ test is required. Notice in Table C that the   nondirectional
$df = N - 2$ are listed down the left side. Assuming a significance
level of .05 and remembering that $N = 42$ (and thus
$df = N - 2 = 40$), we see that the critical value of r is _____ . This   .3044
means that if the observed r is less than or equal to $-.3044$ or greater
than or equal to .3044, then H_0 should be rejected. For this
illustration, $r = -.33$, so H_0 is _____ . It appears that the younger   rejected
the woman, the more feminist her attitude. A formal summary of this
problem is presented in Table 9-3, which you should complete now.

Table 9-3 Guided Computational Example for a Test of the Significance of a Correlation Coefficient

Hypotheses

H_0: _____

H_1: _____ (nondirectional/directional)

Assumptions and Conditions

1. Subjects are _____ and _____ sampled.
2. The population distributions of both X and Y are _____ in form.

Decision Rules (from Table C of Appendix 2 in the text)

Given a significance level of _____ , a _____ test, and $df = N - 2 =$ _____ :

If _____ , _____ .

If _____ , _____ .

Computation

$r_{obs} =$ _____

Decision

_____ H_0 .

The Difference between Two Correlation Coefficients

Suppose a researcher is interested in the relationship between the number of hours per week that children viewed aggressive or violent television programs and the number of physically aggressive acts those children displayed during two weeks of nursery school. The correlation between these two measurements is .46 for a group of 23 boys but only .22 for a sample of 28 girls. Do the sexes really differ in the degree of this relationship? Specifically, what is the probability that the observed difference between .46 and. 22 is simply a consequence of _____ _____ and that in the population there is actually no difference between the two correlations?

    sampling error

    The hypotheses are that the population correlations are equal versus that they are not equal. In symbols, the null hypothesis is _____ and the alternative is _____ . It is assumed that the subjects are _____ and _____ sampled, that the two groups are _____ of each other, and that the population distributions of X and Y for both groups are _____ in form. Both N_1 and N_2 must be large—greater than _____ —for an accurate result, because this statistical test requires the use of the standard normal distribution. For a nondirectional test at the .05 level, Table A in Appendix 2 of the text shows that the critical values of z are _____ , and the decision rules are

_____ , _____ ,

_____ , _____ .

$H_0 : \rho_1 = \rho_2$; $H_1 : \rho_1 \neq \rho_2$
randomly; independently
independent
normal
20

± 1.96;
If $-1.96 < z_{obs} < 1.96$, do not
    reject H_0
If $z_{obs} \leq -1.96$ or $z_{obs} \geq 1.96$,
    reject H_0

    The computation of z_{obs} proceeds as follows. First, an unusual step is taken—the values of r, .46 and .22 in this case, must be transformed into z_r values by using Table D in Appendix 2 of the text. Correlation coefficients are not usually normally distributed, and this transformation normalizes them so that the standard normal distribution can be used. Turn to this table now. Notice that the table lists values of r from .000 to .995 and their corresponding transformed values, z_r. For $r_1 = .46$, $z_{r_1} =$ ____ , and for $r_2 = .22$, $z_{r_2} =$ ____ . Then use the formula

    .497

    .224

$$z_{obs} = \frac{z_{r_1} - z_{r_2}}{\sqrt{\dfrac{1}{N_1 - 3} + \dfrac{1}{N_2 - 3}}}$$

in which N_1 and N_2 are the number of pairs of scores entering into r_1 and r_2, respectively. In this example, $N_1 = \_\_$ and $N_2 = \_\_$. Thus,

23; 28

$$z_{obs} = \underline{\hspace{1.5cm}} = \underline{\hspace{1.5cm}} = \underline{\hspace{1.5cm}}$$

$$\frac{.497 - .224}{\sqrt{\frac{1}{20} + \frac{1}{25}}} = \frac{.273}{.300} = .91$$

This value of z_{obs} does not exceed the z_{crit} of ± 1.96, so H_0 (is/is not) _____ rejected. A formal outline of this problem is presented in Table 9-4, which you should now complete.

is not

Table 9-4 Guided Computational Example for a Test of the Difference between Two Correlation Coefficients

Hypotheses

H_0: _____

H_1: _____ (nondirectional/directional)

Assumptions and Conditions

1. The subjects are _____ and _____ sampled.
2. The two groups are _____ .
3. The population distributions of X and Y for both groups are _____ in form.
4. Both N_1 and N_2 are greater than _____ .

Decision Rules

Given a significance level of _____ and a _____ test:

If _____ , _____ .

If _____ , _____ .

Computation

	Group 1 (Boys)	Group 2 (Girls)
	$N_1 = 23$	$N_2 = 28$
	$r_1 = .46$	$r_2 = .22$
	$z_{r_1} = \_\_\_\_$	$z_{r_2} = \_\_\_\_$

$$z_{obs} = \frac{z_{r_1} - z_{r_2}}{\sqrt{\frac{1}{N_1 - 3} + \frac{1}{N_2 - 3}}} = \underline{\hspace{2cm}} = \underline{\hspace{2cm}}$$

Decision

_____ H_0.

SELF-TEST

1. The formula for testing a single sample mean $[t = (\overline{X} - \mu)/s_{\overline{x}}]$ can be generalized to handle any sample statistic. Express the generalized formula in words.

2. If the sample statistic is the difference between two means, what is the null hypothesis?

3. Why must a different statistical technique be employed when the same subjects are measured twice than when the two groups of measurements are made on different random samples?

4. Under what conditions will the assumption of a normal distribution of $\overline{X}_1 - \overline{X}_2$ be met?

5. In a test of two correlated means, what is the mean of the sampling distribution, assuming the null hypothesis to be true?

6. Why is it necessary to assume in so many of the techniques described in this chapter that the sample distribution is normal in form?

*7. How is it possible to have a significant correlation between two variables for girls and no significant correlation for boys, but no significant difference between the two correlations? What does such a situation tell us about interpreting statistical significance?

*8. Is it possible for there to be a significant difference between two correlated means but no correlation between the two sets of measures on which means were computed? Explain.

*9. What is the relationship between the size of N and the likelihood of obtaining a significant result?

*Questions preceded by an asterisk can be answered on the basis of the discussion in the text, but the discussion in this Study Guide does not answer them.

EXERCISES

Follow the appropriate guided computational example as an outline for each statistical test requested in these exercises. Assume the .05 level of significance except where another is stated, but you will need to decide whether the alternative hypothesis is directional or nondirectional from the information given in the question.

1. Two samples of 21 men each are brought into a room and told they are to help another person in the next room learn a simple problem by giving him an electric shock as a punishment for wrong responses. The subject has a dial with 20 levels of shock that he can administer, 20 being the highest level. Another person is with the subject during the observation; he encourages one group to administer higher and higher levels of shock, whereas he does not so encourage members of the other group. The person supposedly receiving the shocks and the "persuader" are really actors,

but the people administering the "shocks" do not know this. The point of the study is to determine whether people are influenced by pressure from others when given the opportunity to hurt other individuals without fear of disapproval. The hypothetical results are these: The average shock level administered by the subjects who were encouraged to use higher levels was $16.2\left(s_1{}^2 = 12\right)$, and that administered by the other group was $13.8\left(s_2{}^2 = 13\right)$. Evaluate the significance of this difference.[1]

2. Psychologists interested in the vulnerability of eye witnesses to the effects of leading questions showed a film of a traffic accident to a group of college students. The students

---

[1] Adapted from S. Milgram, "Group Pressure and Action Against a Person," *Journal of Abnormal and Social Psychology* 69 (1964): 137-43.

then completed a questionnaire that included the critical question of how fast two automobiles were traveling when they collided. Thirty of the students were asked "About how fast were the cars going when they smashed into each other?" Thirty other students had the same question but with the less violent word "contacted" substituted for "smashed into." The "smashed" group's mean speed estimate was 40.5 with $s_1 = 5.0$. The "contacted" group's mean was 31.0 with $s_2 = 4.5$. Test the significance of the difference between these means.[2]

3. It has been found that leadership, or status, in colonies of male monkeys is related to the level of the male hormone testosterone in the animal's blood: Higher levels correspond to higher status. Such a correlation might indicate that the testosterone level causes the leadership qualities, but the causality could also be in the other direction. As an experiment, 10 male monkeys were reared individually, and their testosterone levels were recorded. Then the 10 males were put together in a large living environment for two months, after which their leadership status and testosterone levels were recorded. The resulting data included testosterone levels before and after the group living experience and the correlations between testosterone levels and leadership. They are shown in the accompanying table.
    If the level of testosterone causes leadership status, then there should be a strong positive correlation between the two testosterone levels (before and after group living) and between each testosterone assessment and the leadership rating. Furthermore, there should be no difference between the average testosterone level before the group living and that after it. On the other hand, if the direction of

causality is reversed and leadership status causes the testosterone levels to change, then there should be no correlation between testosterone level before the group living and leadership, but there should be a significant correlation between testosterone level after the group living and leadership. Statistically evaluate:

a. the correlation between testosterone level before group living and leadership
b. the correlation between testosterone level after group living and leadership
c. the mean difference between testosterone level before the group living and the level after it

What do these results indicate about the direction of causality between testosterone levels and social leadership among monkeys?[3]

| Subject | Testosterone Level | | Leadership |
	Before	After	After
a	5	7	6
b	9	8	7
c	5	5	7
d	6	8	6
e	2	6	4
f	2	6	5
g	7	8	7
h	2	6	4
i	3	11	12
j	4	7	8

$r = .83$

$r = -.10$

4. It is frequently important to ask whether certain personality traits remain consistent as individuals grow up. For example, is a child who is aggressive at age 5 also likely to be aggressive at age 18? The Fels Research

---

[2] Adapted from E. F. Loftus and J. C. Palmer, "Reconstruction of Automobile Destruction: An Example of the Interaction between Language and Memory," *Journal of Verbal Learning and Verbal Behavior* 13 (1974): 585–589.

[3] Based on, but not identical to, studies by I. Bernstein and T. Gordon reported in *Newsweek* (International Edition), 27 August 1973, p. 46.

Institute in Yellow Springs, Ohio, conducted a longitudinal study of children who were seen regularly throughout their childhood in a variety of testing and social contexts, and they were rated for social and intellectual behavior. For 27 males, the correlation between aggressiveness at age 5 and the same variable at age 18 was .39, whereas the correlation for a sample of 32 girls was .21. Assume that a personality theory predicts that aggressiveness should be consistent over age for both sexes.[4]

a.   Are each of these values of r significantly different from zero?

b.   Is the degree of this relationship different for the two sexes?

ANSWERS

Table 9-1. H_0: $\mu_1 = \mu_2$, H_1: $\mu_1 \neq \mu_2$; randomly, independently; independent; homogeneous; $\overline{X}_1 - \overline{X}_2$, normal; .05, nondirectional, 26. If $-2.056 < t_{obs} < 2.056$, do not reject H_0. If $t_{obs} \leq -2.056$ or $t_{obs} \geq 2.056$, reject H_0. $t_{obs} = 2.14$. Reject H_0.

Table 9-2. H_0: $\mu_1 = \mu_2$, H_1: $\mu_2 > \mu_1$; randomly, independently; correlated; normal; .05, directional, 11. If $t_{obs} < 1.796$, do not reject H_0. If $t_{obs} \geq 1.796$, reject H_0. $N = 12$, $\sum D_i = 12$, $\sum D_i^2 = 34$, $\left(\sum D_i\right)^2 = 144$, $t_{obs} = 2.45$. Reject H_0.

Table 9-3. H_0: $\rho = 0$, H_1: $\rho \neq 0$; randomly, independently; normal; .05, nondirectional, 40. If $-.3044 < r_{obs} < .3044$, do not reject H_0. If $r_{obs} \leq -.3044$ or $r_{obs} \geq .3044$, reject H_0. $r_{obs} = -.33$. Reject H_0.

Table 9-4. H_0: $\rho_1 = \rho_2$, H_1: $\rho_1 \neq \rho_2$; randomly, independently; independent; normal; 20; .05, nondirectional. If $-1.96 < z_{obs} < 1.96$, do not reject H_0. If $z_{obs} \leq -1.96$ or $z_{obs} \geq 1.96$, reject H_0. $z_{r_1} = .497$, $z_{r_2} = .224$, $z_{obs} = .91$. Do not reject H_0.

Self-Test. (1) The formula states that a value (e.g., a statistic) minus its population value divided by an estimate of the standard error of that value is distributed as t. **(2)** $\mu_1 = \mu_2$. **(3)** When the groups of measurements are correlated (as when the same or matched subjects contribute a score to each distribution), the standard error is different than when the two distributions are independent, thus a different procedure is required. **(4)** If the two population distributions are normal or if the sample sizes are large regardless of the form of their population. **(5)** 0. **(6)** This permits the use of the percentiles of the standard normal or t distribution. **(7)** For samples of 27 males and 27 females, a correlation of .33 for the girls would be significant but an $r = .31$ for the boys would not be significant; yet there would be no significant difference between .33 and .31. Saying that there is a correlation for girls and no correlation for boys but no difference between them depends on using a dichotomous relationship/no relationship decision. In fact, the probability of each H_0 is a continuum, and the probabilities should be stated. The problem of "accepting the null" must also be recalled here. The nonsignificant correlation cannot be directly translated into the conclusion that *no* correlation exists for boys. **(8)** Yes, because the correlation between two sets of measures is independent of the means. **(9)** As N (or the degrees of freedom) increases, the critical values for many statistical tests decrease, thereby making it more likely that H_0 will be rejected if it is actually false. If H_0 is actually true, the size of N does not influence the probability of a significant result.

---

[4] Based on, but not identical to, studies by J. Kagan and H. A. Moss, *Birth to Maturity* (New York: John Wiley & Sons, 1962.)

Exercises. (1) Follow the format of Table 9-1 for a t test of independent groups. Nondirectional test at .05 with $df = 40$, $t_{crit} = 2.021$, $t_{obs} = 2.20$. Reject H_0. **(2)** Follow the format of Table 9-1 for a t test of independent groups. Nondirectional test at .05 with $df = 58$, $t_{crit} = 2.00$ (approximately), $t_{obs} = 7.74$. Reject H_0. **(3a)** Follow the format of Table 9-3 for the significance of a correlation. Nondirectional test at .05 with $df = 8$, $r_{crit} = \pm.63$, $r_{obs} = -.10$. Do not reject H_0. **(3b)** Follow the same format and r_{crit} as for (3a) but $r_{obs} = .83$. Reject H_0. **(3c)** Follow the format of Table 9-2 for a t test of correlated groups. Nondirectional test at .05 with $df = 9$,

$t_{crit} = \pm2.262$, $t_{obs} = 3.36$. Reject H_0. Since there was a correlation between hormone level and social leadership after but not before the group rearing and since testosterone levels rose after group rearing, social activity may alter hormone level rather than the reverse. **(4a)** Follow the format of Table 9-3 for the significance of a correlation. For males: A directional test at .05 with $df = 25$, $r_{crit} = .3233$, $r_{obs} = .39$. Reject H_0. For females: A directional test at .05 with $df = 30$, $r_{crit} = .2960$, $r_{obs} = .21$. Do not reject H_0. **(4b)** Follow the format of Table 9.4. A nondirectional test at .05, $z_{crit} = \pm1.96$, $z_{r_1} = .412$, $z_{r_2} = .213$, $z_{obs} = .72$. Do not reject H_0.

STATISTICAL PACKAGES

StataQuest

StataQuest can be used to compute a t test for two means based upon two independent samples and for a mean difference based on one sample or from two correlated samples. While StataQuest will compute correlation coefficients (see Chapter 6) and it does test the significance of a Pearson r, it does not test the significance of the difference between two rs (see example, Table 9-4).

Two Independent Means (Table 9-1)

Table 9-1 of the *Guide* gives an example of a t test of two means based upon two independent samples, but the raw data are not presented. To create a new example of the effects of a drug on memory, new raw data are presented below.

Drugged (1)	Non-Drugged (2)
12, 9, 13, 6, 11, 11, 14,	7, 4, 8, 9, 5, 7, 10,
10, 10, 12, 9, 7, 13, 12	6, 6, 8, 9, 7, 7, 6, 9

Start the StataQuest program. Bring up the Stata Editor window by clicking on the *Editor* button at the top of the screen.

Two methods of data entry are possible for this example. In the first, the two groups of memory measures are placed in two separate variables. The first variable is the memory measure for the group that received the drug, the second variable is the measure for the group that did not receive the drug. The number of cases in each group does not have to be equal. Enter the measures from the two groups into the first two columns of the spreadsheet. Label the two columns with the variable names *Drug* and *Nondrug*. To label, double click on the column heading. The data you have entered should now appear on the spreadsheet as follows:

Drug	Nondrug
12	7
9	4
13	8
6	9
11	5
11	7
14	10
10	6
10	6
12	8
9	9
7	7
13	7
12	6
	9

To save the data that you entered:

> *File>Save as*
>> (Enter drive and file name for the save file.)
>> OK

Open a log file to save or print the results of the analysis. Click on the *Log* button at the top left of the screen, enter a name and drive, and click OK.

> To carry out the independent sample *t* test:

> *Statistics>Parametric tests>2-sample t test*
>> Options: 2 independent data variables
>> OK
>>> Data variable #1: Drug   [click to transfer]
>>> Data variable #2: Nondrug
>>> Unequal variances: No
>>> OK

The output consists of a table which gives the number of observations (Obs), mean, and standard deviation for the two groups and for the combined scores. The information at the bottom of the table indicates that the $t = 4.64$, the number of degrees of freedom is 27, and the probability of a *t* this large or larger with a two-tailed test is equal to .0001. StataQuest does not allow for a directional hypothesis. If a one-tailed test is needed, the probability of the observed *t* is exactly one half the value given on the output, as long as the difference between the means is in the hypothesized direction.

The second method of data entry is to code a treatment variable (*Trt*) as to whether or not the group received the drug, and place the scores for all subjects in an outcome variable (*Meas*). To enter data in the spreadsheet, enter the treatment code *Trt* (1=Drugged; 2=Non-Drugged) in the first column, followed by *Meas*, the memory measure, in the second column. Then select the "1 data variable, 1 group variable" option below to carry out the *t* test:

> *Statistics>Parametric tests>2-sample t test*
>> Options: 1 data variable, 1 group variable
>> OK
>>> Data variable: Meas   [click to transfer]
>>> Group var. (2 groups): Trt
>>> Unequal variances: No
>>> OK

This is the method reflected in the accompanying program output. Except for the names of the variables, the results of both methods are identical. Clear the spreadsheet to input data for the next example:

> *File>New*
>> OK

StataQuest Program Output

```
. ttest Meas, by(Trt)

   Variable |       Obs          Mean     Std. Dev.
   ---------+------------------------------------
         1 |        14      10.64286      2.307418
         2 |        15           7.2      1.656157
   ---------+------------------------------------
   combined |        29      8.862069      2.628472

           Ho:   mean(x)  -  mean(y)  =  0   (assuming equal variances)
                        t  =  4.64 with 27 d.f.
               Pr > |t|  =  0.0001
                 95% CI  =  (1.9206171,4.9650971)
```

Two Correlated Means (Table 9-2)

When two means are based on a single group of subjects (for example, pre-posttest scores) or on two correlated samples (for example, parent-child pairs), the means are no longer independent and the independent sample t test is not valid. StataQuest can be used to perform a t test on two correlated means, as long as the data are entered in pairs.

In this example, the two variables are entered as paired test scores, *Test1* and *Test2*. On the Stata Editor spreadsheet, enter the scores for *Test1* in the first column and the corresponding scores for *Test2* in the second column. Label the columns *Test1* and *Test2*. To label a column, point to the column heading and double click. The spreadsheet for the correlated sample t test should appear as follows:

Test1	Test2
5	6
6	7
4	3
4	6
2	2
1	3
2	4
5	4
5	7
5	4
7	9
3	6

To save the data that you entered:

> *File>Save as*
>> (Enter drive and file name for the save file.)
>> OK

To compare the means of the two correlated samples:

> *Statistics>Parametric tests>Paired t test*
>> Data variable #1: Drug   [click to transfer]
>> Data variable #2: Nondrug
>> OK

Note that the example in Table 9-2 is to be a directional test, whereas StataQuest produces output for a nondirectional test. To obtain a nondirectional probability, replace the probability value on the second-to-last line of the output, "$Pr > |t| = 0.0323$" with a value exactly one half the value shown, as long as the difference between the means is in the hypothesized direction.

Print the log file for the above analyses, and exit StataQuest:

> *File>Print Log*
>> OK
> *File>Exit*

```
. ttest Test1 = Test2

  Variable |       Obs          Mean     Std. Dev.
  ---------+----------------------------------------
     Test1 |        12      4.083333      1.78164
     Test2 |        12      5.083333     2.065224
  ---------+----------------------------------------
     diff. |        12            -1     1.414214

            Ho:   mean difference = 0   (paired data)
                        t = -2.45 with 11 d.f.
                  Pr > |t|  = 0.0323
    95% CI for difference = (-1.898548,-.101452)
```

Minitab

Minitab will compute a *t* test for two means based upon two independent samples and for a mean difference based on one sample or from two correlated samples. While Minitab will compute correlation coefficients (see Chapter 6), it does not test the significance of a Pearson *r* or the significance of the difference between two *r*s. These tests can be performed by looking in the tables in your text (see example, Table 9-3) or by making a simple calculation (see example, Table 9-4).

Two Independent Means (Table 9-1)

Table 9-1 of the *Guide* gives an example of a *t* test of two means based upon two independent samples, but the raw data are not presented. So, to create a new example of the effects of a drug on memory, new raw data are presented below.

Drug (1)	Non-Drug (2)
12, 9, 13, 6, 11, 11, 14,	7, 4, 8, 9, 5, 7, 10,
10, 10, 12, 9, 7, 13, 12	6, 6, 8, 9, 7, 7, 6, 9

Two methods of data entry are possible for this example. If you are entering data from the above table, the first method, which is simplest, is to place the two groups of memory measures in two separate columns. In the first column place the memory measure for the group that received drug; the second column contains the measure for the group that did not receive the drug. The number of cases in each group does not have to be equal for the independent sample *t* test.

Start the Minitab program. To enter data on the spreadsheet, click on the Data window. Enter the values from the two groups into the first two columns of the spreadsheet, respectively. Label the two columns with the variable names *Drug* and *Nondrug*. To enter the variable names, type them in at the top of the columns. The data you have entered should now appear on the spreadsheet as follows:

Drug	Nondrug
12	7
9	4
13	8
6	9
11	5
11	7
14	10
10	6
10	6
12	8
9	9
7	7
13	7
12	6
	9

To save the data that you entered:

> *File>Save Worksheet As*
> > (Enter drive and file name for the Minitab save file.)
> > OK

To carry out the independent sample *t* test:

> *Stat>Basic Statistics>2-Sample t*
> > Samples in different columns: Yes
> > > First: Drug   [double-click to transfer]
> > > Second: Nondrug
> > Alternative: not equal
> > Assume equal variances: Yes
> > OK

The accompanying program output, which reflects the first method of data entry, contains more information than you need at this point. For current purposes, the top of the table gives the *N*, mean, and standard deviation for the two groups, and the second line from the bottom of the table indicates that the $t = 4.64$, the probability of a *t* this large or larger with a two-tailed test is equal to .0001, and the number of degrees of freedom is 27. If a one-tailed test is needed, the Alternative subcommand can be changed to "greater than" if the Drug group is expected to have a higher mean than the Nondrug group. If the reverse is true, the alternative should be set to "less than."

The second method of data entry, which Minitab calls the "stacked" method, is to code a treatment variable (*Trt*) to indicate whether or not the group received the drug, and place the memory score in a single outcome variable (*Meas*). In the first column of the spreadsheet enter the treatment code *Trt* (1=Drugged; 2=Non-Drugged). In the second column enter the value for the memory measure *Meas*. The spreadsheet for using the second method of data entry would appear as follows:

Trt	Meas
1	12
1	9
1	13
⋮	⋮
1	12
2	7
2	4
2	8
⋮	⋮
2	9

The *t* test for stacked data entry can be carried out by selecting the "Samples in one column" subcommand in Minitab, as shown below:

Stat>Basic Statistics>2-Sample t
>   Samples in one column: Yes
>   Samples: Meas   [double-click to transfer]
>   Subscripts: Trt
>   Alternative: not equal
>   Assume equal variances: Yes
>   OK

Except for the variable names, the result obtained using either method is the same.
>   To print the output from the Session window:

>   *File>Print Window*
>   Print Range: All
>   OK

To input data for the next worked example, clear the Session window and spreadsheet:

>   *Edit>Select All*
>   *Edit>Delete*
>   *File>New Worksheet*

Minitab Program Output

Two Sample T-Test and Confidence Interval

```
Two sample T for Drug vs Nondrug
            N       Mean      StDev     SE Mean
Drug       14      10.64       2.31       0.62
Nondrug    15       7.20       1.66       0.43


95% CI for mu Drug - mu Nondrug: ( 1.92,   4.97)
T-Test mu Drug = mu Nondrug (vs not =): T= 4.64   P=0.0001   DF=  27
Both use Pooled StDev = 2.00
```

Two Correlated Means (Table 9-2)

When two means are based on a single group of subjects (for example, pre-posttest scores) or on two correlated samples (for example, parent-child pairs), the means are no longer independent and the independent sample *t* test is not valid. Minitab can be used to perform a test on two correlated means, as long as the data are entered in pairs. Then a difference score between the pairs is calculated, and a one-sample *t* test can be run on the mean difference score.

In this example, the two variables are entered as paired test scores, *Test1* and *Test2*. The difference score is placed in a third variable *Diff*, which is obtained by subtracting *Test2* from *Test1*.

In the Data window, enter each score from Table 9-2 for *Test1* and the corresponding score for *Test2* in the first two columns. Label the columns *Test1* and *Test2*. Label the third column by clicking on the blank space below the column designator *C3* and typing in the variable name, *Diff*.

The spreadsheet for the correlated sample *t* test analysis should now appear as follows:

Test1	Test2	Diff
5	6	
6	7	
4	3	
4	6	
2	2	
1	3	
2	4	
5	4	
5	7	
5	4	
7	9	
3	6	

To calculate the difference scores and place them in the variable *Diff*:

> *Calc>Calculator*
> > Store result in variable: Diff   [double-click to transfer]
> > Expression: **Test2 - Test1**   [double-click to transfer, or type in]
> > OK

To save the worksheet to this point:

> *File>Save Worksheet As*
> > (Enter drive and file name for the save file.)
> > OK

To compare the means of the two correlated samples:

> *Stat>Basic Statistics>1-Sample t*
> > Variables: Diff   [double-click to transfer]
> > Test mean: 0
> > Alternative: greater than
> > OK

Note that the example in Table 9-2 is a directional test, which is specified in the command by setting Alternative to "greater than," meaning that Test2 is expected to be larger than Test1 and the *diff* will be "greater than" zero. This is indicated on the first line of the output as "Test of mu = 0 vs mu > 0."

Print the output from the Session window:

File>Print Window
   Print Range: All
   OK

Exit Minitab:

File>Exit

T-Test of the Mean

Test of mu = 0.000 vs mu > 0.000

Variable	N	Mean	StDev	SE Mean	T	P
Diff	12	1.000	1.414	0.408	2.45	0.016

SPSS

SPSS will compute a *t* test for two means from independent samples and a mean difference from one sample or two correlated samples. SPSS *Bivariate* command will compute correlation coefficients (see Chapter 6) and test the significance of a Pearson *r*, but it will not test the significance of the difference between two *r*s (see Table 9-4).

Two Independent Means (Table 9-1)

Table 9-1 of the *Guide* gives an example of a *t* test of two means based upon two independent samples, but the raw data are not presented. So, to create a new example of the effects of a drug on memory for words, new raw data are presented below.

Drugged (1)	**Non-Drugged (2)**
12, 9, 13, 6, 11, 11, 14,	7, 4, 8, 9, 5, 7, 10,
10, 10, 12, 9, 7, 13, 12	6, 6, 8, 9, 7, 7, 6, 9

SPSS requires "stacked" data entry consisting of two variables: a treatment variable with a code for drugged and non-drugged groups (*Trt*), and the measurements of memory (*Meas*).

Start the SPSS program. The Data Editor window should fill the screen. The data are entered by giving the group number (1=Drugged; 2=Non-Drugged) followed by the measurement value, one subject per line. Label the two columns with the variable names *Trt* and *Meas*. To define a variable name for a column, double click on the column heading, then type in the variable name. The spreadsheet should appear as follows:

Trt	Meas
1	12
1	9
1	13
⋮	⋮
1	12
2	7
2	4
2	8
⋮	⋮
2	9

Carry out the *t* test analysis to test for a difference of means:

> *Statistics>Compare Means>Independent-Samples T Test*
> > Test Variable(s): Meas   [highlight and transfer]
> > Grouping Variable: Trt (? ?)
> > Define Groups: Use specified values
> > > Group 1: 1
> > > Group 2: 2
> > > Continue
> > OK

The top table on the output gives the codes you entered for the groups (listed as 1.00 and 2.00) plus the N, mean, standard deviation, and standard error of the mean for the two groups. The bottom table has two rows. The row labeled "Equal variances assumed" gives the results of the t test for the procedure described in your text (ignore the lower row). Note that the probabilities listed for the t test are two-tailed, appearing under the heading "Sig. (2-tailed)." For directional hypotheses, divide the listed p value in half.

The output from this program is wider than normal, and a better result can be obtained if it is printed across the length of the page in Landscape orientation. To print the output:

> *File>Print*
> > Properties>Paper
> > > Orientation: Landscape
> > > OK
> > Print range: All visible output.
> > OK

To clear the output for the next part of the exercise:

> *File>Close*

To save the data that you entered and clear the spreadsheet for new data entry:

> *Window>SPSS Data Editor*
> *File>Save as*
> > (Enter device and file name for SPSS save file.)
> > Save
> *File>New>Data*

SPSS Program Output

T-Test

Group Statistics

	TRT	N	Mean	Std. Deviation	Std. Error Mean
MEAS	1.00	14	10.6429	2.3074	.6167
	2.00	15	7.2000	1.6562	.4276

Independent Samples Test

		Levene's Test for Equality of Variances		t-test for Equality of Means						95% Confidence Interval of the Mean	
		F	Sig.	t	df	Sig. (2-tailed)	Mean Difference	Std. Error Difference		Lower	Upper
MEAS	Equal variances assumed	1.580	.220	4.641	27	.000	3.4429	.7419		1.9206	4.9651
	Equal variances not assumed			4.588	23.469	.000	3.4429	.7504		1.8922	4.9935

Two Correlated Means (Table 9-2)

When the two means are based on a single group of subjects (for example, pre-posttest scores) or on two correlated samples (for example, parent-child pairs), each pair of measures is entered on a separate line on the spreadsheet. In this example, the two variables to be entered are the paired test scores, *Test1* and *Test2*.

In the Data Editor window enter the scores for *Test1* in the first column and the corresponding scores for *Test 2* in the second column. Label the columns *Test1* and *Test2*. To label a column, point to the column heading and double click. The spreadsheet for the correlated sample *t* test should appear as follows:

Test1	Test2
5	6
6	7
4	3
4	6
2	2
1	3
2	4
5	4
5	7
5	4
7	9
3	6

Run the analysis for comparing the means of two correlated samples:

> *Statistics>Compare Means>Paired-Samples T Test*
>> Paired Variables: Test1 Test2   [highlight and transfer]
>> OK

Values of *N*, the mean and standard deviation for each group are reported in the top output table. The correlation between the two tests, and the significance of the correlation coefficient obtained, is reported in the middle output table.

Note that the probability, .032, listed on the bottom output table under "Sig. (2-tailed)" is nondirectional. Since the example in Table 9-2 is directional, the listed probability value must be halved. The mean difference and *t* value are negative because the program subtracts the second variable from the first. Since we hypothesized that the second test score would be higher than the first, this result is consistent with our hypothesis, and it is therefore appropriate to report that the one-tailed probability is .016.

To print the output:

> *File>Print*
>> Print range: All visible output.
>> OK

To save the data that you entered:

> *Window>SPSS Data Editor*
> *File>Save as*
> (Enter device and file name for SPSS save file.)
> Save

Exit SPSS:
> *File>Exit SPSS*

T-Test

Paired Samples Statistics

		Mean	N	Std. Deviation	Std. Error Mean
Pair 1	TEST1	4.0833	12	1.7816	.5143
	TEST2	5.0833	12	2.0652	.5962

Paired Samples Correlations

		N	Correlation	Sig.
Pair 1	TEST1 & TEST2	12	.739	.006

Paired Samples Test

		Paired Differences			95% Confidence Interval of the Difference				Sig. (2-tailed)
		Mean	Std. Deviation	Std. Error Mean	Lower	Upper	t	df	
Pair 1	TEST1 - TEST2	-1.0000	1.4142	.4082	-1.8985	-.1015	-2.449	11	.032

CHAPTER 10

SIMPLE ANALYSIS
OF VARIANCE

CONCEPT GOALS

Be sure that you thoroughly understand the following concepts and how to use them in statistical applications.

- ◆ Notation: X_{ij}, p, n_j, N, T_j, $\overline{X}_j$, T, $\overline{X}$
- ◆ Sum of squares (SS), degrees of freedom (df), mean square (MS)
- ◆ Between-groups and within-groups variance estimates
- ◆ Treatment variability and error variability
- ◆ F distribution and F ratio
- ◆ Assumptions of the analysis of variance

GUIDE TO MAJOR CONCEPTS

In Chapter 9 you learned how to test the hypothesis that the means of two groups of scores differed from each other only because of _____ _____ . But suppose you have more than two groups. What is the probability that the several means of these groups differ from one another by sampling error alone? A problem of this type often requires the statistical technique known as the _____ _____ _____ .

sampling error

analysis of variance

Notation

It will be helpful to review the notation and terminology presented in the text. The data for a simple analysis of variance can be cast into a form similar to that presented in Table 10-1. The scores are clustered into p groups. Each score is designated by X_{ij}, meaning the ith subject in the jth group. So X_{42} is the _____ subject in the _____ group. The symbol n_j denotes the number of subjects in the jth group: Thus n_2 is the number of subjects in the _____ group. N stands for the total number of subjects in the entire data set: $n_1 + n_2 + \cdots + n_p = $ ___ . There are p groups, so the score for the third subject in the last group is _____ , and the number of subjects in the last group is _____ . Sometimes it is convenient to talk about some subject in some group without specifying any particular one. In this case, one uses the subscripts i and j rather than specific numbers; the score of some unspecified subject is thus denoted _____ , and the number of subjects in some unspecified group can be expressed by _____ .

fourth

second

second

N

X_{3p}

n_p

X_{ij}

n_j

The mean of the jth group is written $\overline{X}_j$; the mean of the first group is thus _____ . The mean over all subjects in the data set, called the **grand mean**, is written without subscripts, simply as _____ . Notice that the formula for the grand mean involves two summation signs: $\sum_{j=1}^{p}\sum_{i=1}^{n_j} X_{ij}/N$. This is a symbolic way of directing you to sum the scores of all the subjects within each group $\left(\sum_{i=1}^{n_j}\right)$ and over all groups $\left(\sum_{j=1}^{p}\right)$. One must also be careful to note which subscript, i or j, is being summed. For example, the sum of all the

$\overline{X}_1$

$\overline{X}$

Table 10-1 General Notation for Simple Analysis of Variance

	Group 1	Group 2	$\cdots$	Group p	
	X_{11}	X_{12}	$\cdots$	X_{1p}	
	X_{21}	X_{22}	$\cdots$	X_{2p}	
	X_{31}	X_{32}	$\cdots$	X_{3p}	
	X_{41}	X_{42}	$\cdots$	X_{4p}	
	.	.	.	.	
	.	.	.	.	
	.	.	.	.	
	$X_{n_1 1}$	$X_{n_2 2}$	$\cdots$	$X_{n_p p}$	
					Grand Total
Group Totals	T_1	T_2	$\cdots$	T_p	T
					Grand Mean
Group Means	$\overline{X}_1 = \dfrac{T_1}{n_1}$	$\overline{X}_2 = \dfrac{T_2}{n_2}$	$\cdots$	$\overline{X}_p = \dfrac{T_p}{n_p}$	$\overline{X}$

X_{ij} = score for the ith subject in the jth group

p = number of groups

n_j = number of subjects in the jth group

$$N = n_1 + n_2 + \cdots + n_p = \sum_{j=1}^{p} n_j = \text{total number of subjects over all groups}$$

T_j = total of all scores in group j

$\overline{X}_j$ = mean for group $j = \dfrac{T_j}{n_j}$

T = grand total over all subjects

$\overline{X}$ = grand mean $= \dfrac{T}{N}$

scores in the second group is written _____ , whereas the $\displaystyle\sum_{i=1}^{n_2} X_{i2}$

sum of the p group means would be designated _____ . By $\displaystyle\sum_{j=1}^{p} \overline{X}_j$

analogy, the formula for the mean of the third group would be

_____ . $\dfrac{\displaystyle\sum_{i=1}^{n_3} X_{i3}}{n_3}$

The analysis of variance also uses some special terminology for discussing variances. In the formula for a sample variance, $s^2 = \dfrac{\sum(X_i - \overline{X})^2}{N-1}$, the numerator is **a sum of squares (SS)**—the sum of the squared deviations of scores about their mean. In the context of the analysis of variance, the sum of these squared deviations is called a _____ _____ _____ , abbreviated _____ . The denominator of the formula, $N-1$, is the number of **degrees of freedom (df)** corresponding to that *SS*. Therefore, in the formula for the variance, one divides the _____ ____ _____ by its _____ ___ _____ . Finally, variance estimates are called **mean squares (MS),** because they are roughly the mean of the squared deviations. So in the context of the analysis of variance, a variance estimate consists of a_____ ____ _____ divided by its _____ _____ _____ ; this quantity is called a _____ _____ . In symbols, _____ = _____ .

> sum of squares
> *SS*
>
> sum
> of squares; degrees of freedom
>
> sum
> of squares; degrees of freedom
> mean square
> $$MS = \frac{SS}{df}$$

Rationale

Now let us consider the rationale of simple analysis of variance. Two estimates of the population variance are needed, one based upon the deviation of scores about their own group mean (the **within-groups mean square**), and the other based upon the deviations of the group means about the grand mean (the **between-groups mean square**). Thus, the basic strategy of the analysis of variance is to derive two variance estimates, or _____ _____ , from the data—one that is sensitive to differences between groups, called the _____ mean square, and one that is not sensitive to group differences, called the _____ mean square.

> mean squares
>
> between-groups
> within-groups

Suppose a scientist is interested in the conservative or liberal character of political opinions of people in four age groups — 20–29, 30–39, 40–49, and 50–59. A questionnaire provides a measure of political opinion. First, consider why two subjects in *different age groups* do not score the same value. Obviously, political opinion may vary with the difference between groups—age. This is called variability due to **treatment effects**. But it is also true that political opinion is not completely determined by the treatment group; subjects will have different political attitudes even if they are the same age. Moreover, the measurement itself is not precisely the same from subject to subject. These other sources of variability are collectively called **error**. The mean square that reflects both variability due to _____ _____ and variability due to _____ is the _____ mean square.

> treatment effects
> error; between-groups

Second, consider why subjects *within a single age group* differ
from one another. Since these subjects all belong to the same age
group, their scores do not vary because of _____   treatment
effects. Variation within a group is caused only by _____ , and   error
such variability is reflected in the _____ mean square.   within-groups
    To summarize, the between-groups mean square reflects
variability associated with _____ _____ and with   treatment effects
subject and measurement _____ , whereas the within-groups   error
mean square reflects variability associated only with _____ .   error
But the null hypothesis for the analysis of variance states that there
are no group differences—no treatment effects—in the population.
According to the null hypothesis, both the between-groups and the
within-groups mean square reflect variability associated only with
_____ . Consequently, under the null hypothesis the two mean   error
squares should be equal except for _____ _____ .   sampling error
    The ratio of two independent variance estimates is a statistic
called ____ . In the case of the analysis of variance, this is expressed: F

$$F_{obs} = \frac{MS_{between}}{MS_{within}}$$

Under the null hypothesis, the value of F_{obs} will vary only due to
sampling error. The F distribution presented in Table E in Appendix
II of your text states how much variability due to sampling error
should occur in the value of F_{obs} if $\mu_1 = \mu_2 = \cdots = \mu_p = \mu$, that is,
under the _____ _____ . If it is not too improbable   null hypothesis
that the actual value of F_{obs} is a product only of sampling error, then
_____ _____ _____ _____ . Since group differences   do not reject H_0
influence the size of the _____ mean square but not the   between-groups
_____ mean square, as group differences increase, the   within-groups
value of F_{obs} will become (larger/smaller) _____ . When the   larger
value of F_{obs} is too large to be explained on the basis of
_____ _____ alone, one _____ _____ .   sampling error; rejects H_0
    Consider the example of age and political opinion mentioned
above. Table 10-2 lists the hypothetical political opinion scores of
five subjects in each of the four age groups. The column means ($\overline{X}_j$)
and the grand mean ($\overline{X}$) are given. The sum of squares between
groups ($SS_{between}$) is the number of subjects in each group ($n_j = 5$)
times the sum of squared differences between each group mean and
the grand mean. For the first group (ages 20–29) compute
$n_1(\overline{X}_1 - \overline{X})^2 =$ _____ $=$ _____ . Adding this quantity for all $5(4-5)^2 = 5$
groups gives $SS_{between} =$ ___ . Now, within each group determine the   10
sum of squared deviations (SS), that is, the sum of squared
differences between each score and its group mean. You should have

Table 10-2 How $MS_{between}$ Is Changed by a Treatment Effect

	Age Groups				
	20–29	30–39	40–49	50–59	
	7	2	9	6	
	5	6	3	2	
	3	9	5	11	
	4	9	6	1	
	1	4	2	5	
Total: $\sum_{i=1}^{n_j} X_{ij}$	20	30	25	25	Grand Total: $\sum_{j=1}^{p}\sum_{i=1}^{n_j} X_{ij} = 100$
Mean: $\overline{X}_j$	4	6	5	5	Grand Mean: $\overline{X} = 5.0$

$SS_{within}: \sum_{i=1}^{n_j}\left(X_{ij} - \overline{X}_j\right)^2$

$SS_{within} = \sum_{j=1}^{p}\sum_{i=1}^{n_j}\left(X_{ij} - \overline{X}_j\right)^2 = \underline{\hspace{1cm}}$

$SS_{between} = n_j \sum_{j=1}^{p}\left(\overline{X}_j - \overline{X}\right)^2 = \underline{\hspace{1cm}}$

$MS_{between} = \dfrac{SS_{between}}{df_{between}} = \dfrac{SS_{between}}{p-1} = \underline{\hspace{1cm}} = \underline{\hspace{1cm}}$

$F_{obs} = \dfrac{MS_{between}}{MS_{within}} = \underline{\hspace{1cm}}$

$MS_{within} = \dfrac{SS_{within}}{df_{within}} = \dfrac{SS_{within}}{N-p} = \underline{\hspace{1cm}} = \underline{\hspace{1cm}}$

$= \underline{\hspace{1cm}}$

Treatments	Age Groups				
	20–29	30–39	40–49	50–59	
	+0	+2	+4	+6	
	7	4	13	12	
	5	8	7	8	
	3	11	9	17	
	4	11	10	7	
	1	6	6	11	
Total: $\sum_{i=1}^{n_j} X_{ij}$					Grand Total: $\sum_{j=1}^{p}\sum_{i=1}^{n_j} X_{ij} = 160$
Mean: $\overline{X}_j$					Grand Mean: $\overline{X} = 8.0$

$SS_{within}: \sum_{i=1}^{n_j}\left(X_{ij} - \overline{X}_j\right)^2$

$SS_{within} = \sum_{j=1}^{p}\sum_{i=1}^{n_j}\left(X_{ij} - \overline{X}_j\right)^2 = \underline{\hspace{1cm}}$

$SS_{between} = n_j \sum_{j=1}^{p}\left(\overline{X}_j - \overline{X}\right)^2 = \underline{\hspace{1cm}}$

$MS_{between} = \dfrac{SS_{between}}{df_{between}} = \dfrac{SS_{between}}{p-1} = \underline{\hspace{1cm}} = \underline{\hspace{1cm}}$

$F_{obs} = \dfrac{MS_{between}}{MS_{within}} = \underline{\hspace{1cm}}$

$MS_{within} = \dfrac{SS_{within}}{df_{within}} = \dfrac{SS_{within}}{N-p} = \underline{\hspace{1cm}} = \underline{\hspace{1cm}}$

$= \underline{\hspace{1cm}}$

found the SS_{within} for the four age groups to be ___ , ___ , ___ and ___ .   20; 38; 30; 62

Adding these gives the sum of squares within groups (SS_{within}),
which in this case equals ____ . To determine the mean square,   150

divide the _____ ____ _____ by its _____   sum of squares; degrees

____ _____ . Since $df_{between} = p - 1 =$ _____ and   of freedom; 3

$df_{within} = N - p =$ _____ , $MS_{between} =$ _____ = _____ ,   16; 10/3 = 3.33

$MS_{within} =$ _____ = _____ , and their ratio $F_{obs} =$ _____ .   150/16 = 9.38; .36

Now let us consider what happens to $MS_{between}$ and MS_{within} if
the null hypothesis, $\mu_1 = \mu_2 = \cdots = \mu_p = \mu$, is not true. The scores in
the bottom of Table 10-2 are the same as those at the top, except that
0, 2, 4, or 6 has been added to all the scores in a column to simulate
treatment differences between age groups in the population. Now
compute MS_{within}, $MS_{between}$, and F_{obs} at the bottom of the table in
the same manner as above. You should have found that while
MS_{within} retains the value _____ , $MS_{between}$ now equals _____ .   9.38; 43.33

Thus, differences between groups are reflected in the value of
_____ but not in the value of _____ . F_{obs} now equals $MS_{between}$; MS_{within}

_____ .   4.62

The critical values of the theoretical sampling distribution of the
ratio of two independent variance estimates, called ___ , are presented   F

in Table E in Appendix 2 of your text. Turn to it now. The columns
of the table correspond to the number of _____ _____   degrees of

_____ for the mean square in the _____ of the F ratio,   freedom; numerator

while the rows correspond to the number of _____ _____   degrees of

_____ for the mean square in the _____ of this   freedom; denominator

fraction. For this particular example, the df for $MS_{between}$ is

$p - 1 =$ ___ , while MS_{within} has $df = N - p =$ _____ . The values in   3; 16

the table at the intersection of the column for 3 df and the row for 16
df are ___ and _____ . These represent the critical values of F for a   3.24; 5.29

test at the .05 and the .01 _____ levels, respectively. Thus, the   significance

critical value of F at the .05 level is _____ . Since $F_{obs} \geq F_{crit}$ at   3.24

the bottom of Table 10-2, _____ H_0 .   reject

Computation

The example above used definitional formulas for calculations. This
approach helps us understand how the analysis of variance works, but
other formulas that give the same numerical result are much more
convenient for most applications. Suppose a study of how parents
discipline their 10-year-old children identifies three groups of
parents. One group is very controlling, restrictive, and strict. Another
is quite permissive and laissez-faire, while a third group is between
these extremes, combining firmly enforced rules with provisions for

independence. The children are then observed in a special situation in which they are given 10 opportunities to obey an adult. The number of times the children are obedient is recorded.[1] A guided computational example of this study is given in Tables 10-3 and 10-4.

The null hypothesis that all the population means are equal can be stated in symbols as H_0: _____ = _____ = ... = ____ = _____ .

$$\mu_1 = \mu_2 = \cdots = \mu_p = \mu$$

The alternative is best expressed as the negation of H_0. Thus, H_1: _____ . This is stated below. The assumptions are that the subjects

not H_0

are _____ and _____ sampled, the groups are _____ of one another, the variances in the populations are the same from group to group (i.e., they are _____), and the population distributions are _____ in form. As found in Table E in Appendix 2 of your text, for $df = 2$ and 13 at the .05 level, F_{crit} = _____ . The decision rules are stated as if a one-tailed test is being performed because only large F_{obs} values will lead to rejection of H_0. However, the alternative hypothesis is not considered directional because the direction of differences between

randomly; independently
independent
homogeneous
normal

3.80

Table 10-3 Guided Computational Example: Formal Summary

Hypotheses

    H_0: _____

    H_1: _____ (non-directional)

Assumptions and Conditions

1.   The subjects are _____ and _____ sampled.
2.   The groups are _____ .
3.   The population variances are _____ .
4.   The population distributions are _____ in form.

Decision Rules (from Table E of Appendix 2 in the text)

Given a significance level of _____ and $df = p - 1 =$ _____ and $N - p =$ _____ :

If _____ , _____ .

If _____ , _____ .

Computation

See Table 10-4.

$F_{obs} =$ _____

Decision

_____ H_0.

---

[1] Inspired by, but not identical to, research reported by D. Baumrind, "Current Patterns of Parental Authority," *Developmental Psychology Monographs*, vol. 4. no. 1. pt. 2 (1971).

means is not specified. All tests in the analysis of variance in this text will be nondirectional. Thus, if F_{obs} is greater than or equal to F_{crit}, H_0 is rejected. Formally:

If _____ , _____ .

If _____ , _____ .

$F_{obs} < 3.80$, do not reject H_0

$F_{obs} \geq 3.80$, reject H_0

The computational routine is presented in Table 10-4. For each group, first total the scores to obtain _____ , which when divided by n_j gives the group _____ . Next, sum each squared score within a group to obtain _____ , and then square the group total and divide by n_j to compute _____ . At the right, add up these values over the groups. Then calculate the three intermediate quantities in part B of Table 10-4. Notice that these intermediate quantities are defined differently than they were in the regression and correlation chapters. Then calculate the SS, df, MS, and F_{obs} in the summary table of part C. The numerical results are given at the end of this chapter.

T_j

mean

$\sum_{i=1}^{n_j} X_{ij}^2$

$\dfrac{T_j^2}{n_j}$

Table 10-4 Guided Computational Example: Calculations

A. Data	Discipline Group			
	Controlling	Combination	Permissive	
	3	6	4	
	5	9	7	
	2	5	3	
	1	8	5	
	4	7	6	
			5	
Totals, T_j				$T = \sum_{j=1}^{p} T_j =$ _____
n_j				$N = \sum_{j=1}^{p} n_j =$ _____
Group Means, $\overline{X}_j$				
Sum of squared scores, $\sum_{i=1}^{n_j} X_{ij}^2$				$\sum_{j=1}^{p}\left(\sum_{i=1}^{n_j} X_{ij}^2\right) =$ _____
Squared sum of scores divided by n_j, $\dfrac{T_j^2}{n_j}$				$\sum_{j=1}^{p}\left(\dfrac{T_j^2}{n_j}\right) =$ _____

B. Intermediate Quantities

$$(\text{I}) = \frac{T^2}{N} = \underline{\hspace{2cm}} \qquad (\text{II}) = \sum_{j=1}^{p}\left(\sum_{i=1}^{n_j} X_{ij}^2\right) = \underline{\hspace{2cm}} \qquad (\text{III}) = \sum_{j=1}^{p}\left(\frac{T_j^2}{n_j}\right) = \underline{\hspace{2cm}}$$

C. Summary Table

Source	df	SS	MS	F_{obs}
Between groups	$p - 1 = \underline{\hspace{1cm}}$	$(\text{III}) - (\text{I}) = \underline{\hspace{1cm}}$	$\dfrac{SS_{between}}{df} = \underline{\hspace{1cm}}$	$\dfrac{MS_{between}}{MS_{within}} = \underline{\hspace{1cm}}$
Within groups	$N - p = \underline{\hspace{1cm}}$	$(\text{II}) - (\text{III}) = \underline{\hspace{1cm}}$	$\dfrac{SS_{within}}{df} = \underline{\hspace{1cm}}$	
Total	$N - 1 = \underline{\hspace{1cm}}$	$(\text{II}) - (\text{I}) = \underline{\hspace{1cm}}$		

SELF-TEST

*1. Explain why the analysis of variance is preferable to a series of t tests of the differences between pairs of means as a method of evaluating H_0: $\mu_1 = \mu_2 = \cdots = \mu_p = \mu$.

2. Define the F ratio (as used in this type of analysis of variance). Does the ratio increase or decrease if the values within each group become more variable?

*3. Explain how the partition of variability in the analysis of variance is analogous to the partition of variability in regression and correlation.

4. In the context of an experiment, $MS_{between}$ is sometimes referred to as $MS_{treatment}$ while MS_{within} is sometimes referred to as MS_{error}. Explain.

5. What are the assumptions of the analysis of variance, and why is each necessary?

*Questions preceded by an asterisk can be answered on the basis of the discussion in the text, but the discussion in this Study Guide does not answer them.

EXERCISES

1. A developmental psychologist was interested in testing the popular notion that breast-feeding one's infant increases the warmth and intimacy of the mother-child relationship and serves as a good foundation for this relationship for years to come. The psychologist sent observers to the homes of a sample of three-year-old children who had been breast-fed for 0, 1–2, 2–5, or more than 5 months. These observers spent an entire day in the home and rated the degree of closeness and intimacy in the relationship between mother and child. The data are given in the accompanying table; a high score means a close mother-child relationship. Evaluate these hypothetical data by following the procedures outlined in Tables 10-3 and 10-4.

Months of Breast Feeding			
0	1–2	2–5	5+
5	7	1	4
7	6	3	2
2	3	6	7
3	9	5	4
4	7	6	
2			

2. Various claims have been made for the merits of different kinds of nursery schools. One of the most frequently heard arguments is that less structured schools foster more creativity and intellectual curiosity than traditional, structured programs. To test these claims, an educator located three nursery schools that received children from approximately equal educational and socioeconomic levels of the community. The schools were a traditional school, a Montessori school, and a Summer-hill-type "free" school. The children in these schools were given a nonverbal creativity test. The data are presented in the accom-panying table. Statistically evaluate the observed differences between these schools.[2]

Creativity		
Traditional	Montessori	Free
6	2	1
9	5	3
7	4	3
5	4	5
8	6	
8		

ANSWERS

Table 10-4. $T_j = 15, 35, 30$; $T = 80$; $n_j = 5, 5, 6$; $N = 16$; $\sum X_j^2 = 55, 255, 160$; $\sum X^2 = 470$; $T_j^2/n_j = 45, 245, 150$; $T^2/N = 440$; **(I)** $= 400$, **(II)** $= 470$, **(III)** $= 440$; between groups: 2, 40, 20; within groups: 13, 30, 2.3077; Total: 15, 70; $F_{obs} = 8.67**$.

Self-Test. (1) The statistical question addressed by the analysis of variance has to do with the set of group means, not with pairs of means; the probability estimate based on numerous t tests would be ambiguous because the several tests would not be independent.
(2) $F = MS_{between}/MS_{within}$, decrease. **(3)** See the text, pages 276–278. In regression and correlation, one partitions the variability of the predicted variable into a portion associated with the predictor variable and a portion not so associated (i.e., error). In the analysis of variance, one partitions the variability in the dependent variable into a portion associated with group membership (i.e., with the independent variable) and a portion associated with error. **(4)** In an experiment, different groups receive different treatments (for example, different drug levels). Differences between the group means are due to differential effects of the treatments plus uncontrolled random effects. Differences within groups are due only to uncontrolled random effects, i.e., error. **(5)** See pages 285–287, of the text.

Exercises. (1) Follow the format given in Tables 10-3 and 10-4. **(I)** $= 432.4500$, **(II)** $= 523$, **(III)** $= 453.4167$; $df_{between} = 3$, $SS_{between} = 20.9667$, $MS_{between} = 6.9889$, $df_{within} = 16$, $SS_{within} = 69.5833$, $MS_{within} = 4.3490$, $df_{total} = 19$, $SS_{total} = 90.5500$; $F_{obs} = 1.61$, F_{crit} (significance level .05, $df = 3$, 16) $= 3.24$, do not reject H_0. **(2) (I)** $= 385.0667$, **(II)** $= 460$, **(III)** $= 432.3667$; $df_{between} = 2$, $SS_{between} = 47.3000$, $MS_{between} = 23.6500$, $df_{within} = 12$, $SS_{within} = 27.6333$, $MS_{within} = 2.3028$, $df_{total} = 14$, $SS_{total} = 74.9333$; $F_{obs} = 10.27$, F_{crit} (significance level .05, $df = 2$, 12) $= 3.88$, reject H_0.

[2] Based on, but not identical to, a study by A. S. Dreyer and D. Rigler, "Cognitive Performance in Montessori and Nursery School Children," *Journal of Educational Research* 62 (1969): 411–16.

STATISTICAL PACKAGES

StataQuest
(To accompany the guided computational example in Table 10-4.)

StataQuest will compute the simple analysis of variance given in Table 10-4. As for the *t* test, Chapter 9, two methods of data entry are possible.

In the first method of data entry, the outcome scores are placed in each of three columns on the spreadsheet, corresponding to each of the three levels of discipline. The number of measures in each group does not have to be equal. The purpose of the analysis is to compare the means of the three groups. Start the StataQuest program. Bring up the Stata Editor window. Enter the data and label the three columns with variable names *Control*, *Combin*, and *Permiss*, corresponding to the three levels. The spreadsheet should appear as follows:

Control	Combin	Permiss
3	6	4
5	9	7
2	5	3
1	8	5
4	7	6
		5

To save the data that you entered:

> *File>Save as*
> > (Enter drive and file name for the save file.)
> > OK

Open a log file to save or print the results of the analysis. Click on the *Log* button at the top left of the screen, enter a name and drive, and click OK.

This analysis is carried out using the "k data variables, one per treatment" option below:

> *Statistics>ANOVA>One-way*
> > Options: k data variables, one per treatment
> > > Data variables: Control Combin Permiss   [click to transfer]
> > > OK
> > > One-way ANOVA plots: Error bar, conf. level *95*
> > > Run

In the second method, you identify the outcome variable, which is the measure of obedience, and the treatment variable, also called the *factor*. In this problem, the factor is the level of discipline (1=Control, 2=Combination, 3=Permissive). Enter the level of discipline, *Discip*, as 1, 2, or 3 in the first column of the spreadsheet, and the corresponding measure of obedience, *Obed*, in the second column. Label the two columns with the variable names *Discip* and *Obed*. The data in the spreadsheet should appear as follows:

Discip	Obed
1	3
1	5
1	2
1	1
1	4
2	6
2	9
2	5
2	8
2	7
3	4
3	7
3	3
3	5
3	6
3	5

Carry out this analysis using the following commands:

Statistics>ANOVA>One-way
        Options: 1 data variable, 1 group variable
                Data variable: Obed   [click to transfer]
                Group var. (3 groups): Discip
                OK
                One-way ANOVA plots: Error bar, conf. level *95*
                Run

The output which reflects the second method of data entry, includes the mean and standard deviation for each group, the one-factor analysis of variance, and a plot of group means and error bars which are based upon the individual group standard deviation, not a pooled estimate.
Except for the variable names, the results using this method of data entry are identical.
    Print both the log file and the graph:

File>Print Log
        OK

File>Print Graph
        OK

Exit StataQuest:

File>Exit

StataQuest Program Output

```
. oneway Obed Discip, tabulate

                         Summary of Obed
          Discip|      Mean    Std. Dev.        Freq.
     ------------+-------------------------------------
             1 |        3     1.5811388            5
             2 |        7     1.5811388            5
             3 |        5     1.4142136            6
     ------------+-------------------------------------
         Total |        5     2.1602469           16

                         Analysis of Variance
       Source           SS          df      MS            F      Prob > F
     ------------------------------------------------------------------------
     Between groups     40.00        2        20.00      8.67     0.0041
      Within groups     30.00       13     2.30769231
     ------------------------------------------------------------------------
         Total          70.00       15     4.66666667

Bartlett's test for equal variances:   chi2(2) =    0.0681  Prob>chi2 =
0.967
```

Minitab
(To accompany the guided computational example in Table 10-4.)

Minitab will compute a simple one-way analysis of variance for independent groups on data such as that presented in Table 10-4. The purpose of the analysis is to compare the means of the three groups being measured. Minitab will compute the mean and standard deviation of each group, carry out a one-factor analysis of variance, and plot group means and error bars. Two methods of data entry are possible in Minitab, "stacked" and "unstacked."

To enter data directly from Table 10-4, the simplest method is to place the outcome data into three columns, each of which corresponds to a discipline group. This is the method Minitab refers to as "unstacked" data entry.

Start the Minitab program. Click on the Data window. Enter the values for the three discipline groups into the first three columns of the spreadsheet. Label these columns *Control*, *Combin,* and *Permiss*. The spreadsheet should now appear as follows:

Control	Combin	Permiss
3	6	4
5	9	7
2	5	3
1	8	5
4	7	6
		5

To save the data as a Minitab worksheet:

> *File>Save Worksheet As*
> > (Enter drive and name for the file.)
> > OK

To carry out the one factor analysis of variance, use the Oneway Unstacked command:

> *Stat>ANOVA>Oneway [Unstacked]*
> > Responses (in separate columns):
> > > Control Combin Permiss   [double-click each to transfer]
> > OK

The output gives more information than we currently need, but it includes the entire analysis of variance table as well as group means and standard deviations. Notice that confidence intervals for each mean are based upon the standard deviation that is pooled across all groups, not the individual standard deviation for each group.

The same result can be produced using the "stacked" method of data entry, in which a treatment variable, or factor, is used to indicate group membership. To enter data for the analysis using the "stacked" method of data entry, enter the factor, *Discip* (coded 1=Control, 2=Combination, 3=Permissive), in one column and the dependent variable, *Obed*, in a second column. Label the two columns with the variable names. The data in the spreadsheet should appear as follows:

Discip	Obed
1	3
1	5
1	2
1	1
1	4
2	6
2	9
2	5
2	8
2	7
3	4
3	7
3	3
3	5
3	6
3	5

Then the analysis may be carried out using the Oneway command:

> *Stat>ANOVA>Oneway*
> Response: Obed   [double-click to transfer]
> Factor: Discip
> OK

To print the results of the Analysis of Variance, print the Session window:

> *File>Print Window*
> Print Range: All
> OK

Exit Minitab:

> *File>Exit*

Minitab Program Output

One-Way Analysis of Variance

```
Analysis of Variance
Source      DF          SS          MS          F          P
Factor       2       40.00       20.00       8.67      0.004
Error       13       30.00        2.31
Total       15       70.00
```

```
                                    Individual 95% CIs For Mean
                                    Based on Pooled StDev
Level       N       Mean       StDev   ---+---------+---------+---------+---
Control     5      3.000      1.581   (------*------)
Combin      5      7.000      1.581                          (------*------)
Permiss     6      5.000      1.414              (------*------)
                                        ---+---------+---------+---------+---
Pooled StDev =     1.519                2.0       4.0       6.0       8.0
```

SPSS

(To accompany the guided computational example in Table 10-4.)

To have SPSS compute the simple analysis of variance given in Table 10-4, you need to identify the independent variable, also known as the treatment variable, and the dependent variable or outcome measure. The treatment variable is called a *factor*, and the simple analysis of variance becomes a one-factor design. In this example the factor is the level of discipline (1=Control, 2=Combination, 3=Permissive); the dependent variable is the measure of obedience. The purpose of the analysis is to compare the means of the three groups being measured.

Start the SPSS program. The Data Editor window should fill the screen.

Enter the level of discipline, *Discip*, in the first column of the spreadsheet, and the corresponding measure of obedience, *Obed*, in the second column. Label the columns with the variable names. To enter a variable name, point to the column heading and double click. The data in the spreadsheet should appear as follows:

Discip	Obed
1	3
1	5
1	2
1	1
1	4
2	6
2	9
2	5
2	8
2	7
3	4
3	7
3	3
3	5
3	6
3	5

To carry out the simple (one-factor) analysis of variance:

> *Statistics>Compare Means>One-Way ANOVA*
> > Dependent List: Obed   [highlight and transfer]
> > Factor: Discip
> > Options:
> > > Statistics: Descriptive
> > > Continue
> > OK

The output includes the group means and standard deviations and the ANOVA table. The group means table requires landscape orientation to fit easily across a page.

To print the output:

> *File>Print*
>> Properties: Paper:
>>> Orientation: Landscape
>>> OK
>> Print range: All visible output.
>> OK

To save the data that you entered:

> *Window>SPSS Data Editor*
> *File>Save as*
>> (Enter device and file name for SPSS save file.)
>> Save

Exit SPSS:

> *File>Exit SPSS*

SPSS Program Output

Oneway

Descriptives

			N	Mean	Std. Deviation	Std. Error	95% Confidence Interval for Mean		Minimum	Maximum
							Lower Bound	Upper Bound		
OBED	DISCIP	1.00	5	3.0000	1.5811	.7071	1.0368	4.9632	1.00	5.00
		2.00	5	7.0000	1.5811	.7071	5.0368	8.9632	5.00	9.00
		3.00	6	5.0000	1.4142	.5774	3.5159	6.4841	3.00	7.00
		Total	16	5.0000	2.1602	.5401	3.8489	6.1511	1.00	9.00

ANOVA

		Sum of Squares	df	Mean Square	F	Sig.
OBED	Between Groups	40.000	2	20.000	8.667	.004
	Within Groups	30.000	13	2.308		
	Total	70.000	15			

PART 3

SPECIAL
TOPICS

CHAPTER 11

INTRODUCTION TO RESEARCH DESIGN

CONCEPT GOALS

Be sure that you thoroughly understand the following concepts and how to use them.

- Steps in conducting research

- Experimental and observational research, causal and noncausal relationships

- Independent and dependent variables

- Operational definitions

- Reliability and validity

- Control groups, extraneous variables, confounding

- Subject and observer bias, single and double blind designs

GUIDE TO MAJOR CONCEPTS

Research design consists of the methods scientists use to make observations that will produce **empirical** information, usually about the relationship between two or more things called **variables**. Science relies heavily on systematic observation, which is why science is called _____ . Typically, researchers observe the relationship between two or more _____ according to systematic methods which collectively are called _____ _____ .

empirical
variables
research
design

The "things" scientists study are _____ , and there are two general kinds, **independent** and **dependent**. The variable thought to influence the other is the _____ variable, while the variable thought to be influenced is the _____ variable. The methods of _____ _____ then, describe the _____ relationship between the _____ and _____ _____ .

variables

independent
dependent
research design
empirical; independent
dependent variables

Such relationships may be **causal** or **noncausal**. For example, the independent variable might actually influence or determine the value of the dependent variable in some direct way. In this case, the relationship is _____ . Alternatively, the two variables might be related, but the _____ variable does not directly produce the values of the _____ variable, in which case the relationship is _____ .

causal
independent
dependent
noncausal

Observational and **experimental** research methods differ in the kind of relationship that can be described. If the purpose of the research is to describe a relationship between two or more variables, regardless of whether that relationship is _____ or _____ , then _____ methods might be used. However, if the purpose of the research is to describe a _____ relationship, then _____ methods would be preferred.

causal; noncausal
observational
causal
experimental

Regardless of which approach is taken, the process of conducting research typically consists of five steps. The first is to **formulate a scientific question**. Because science relies on systematic observations, that is, it is _____ , a proper scientific question usually involves a relationship between at least two observable _____ . Therefore, the first step in conducting research is to _____ __ _____ _____ about two observable variables.

empirical

variables
formulate a scientific question

The next step is to **operationalize** the variables and their relationship. When a variable is defined in terms of the actions required to measure it, the result is an _____ definition.

operational

The third step is **data collection**. Making the actual observations and recording the measurements constitutes _____ _____ .

data collection

The next step is to **analyze the data**. It is in this phase that the numerical information is described, relationships quantified, and inferences drawn. Typically, statistics are used to _____ _____ _____ .

analyze
the data

The fifth step consists of **drawing conclusions and interpretations**. As a result of the data analysis, the researcher attempts to answer the scientific question by _____ _____ _____ _____ .

drawing
conclusions and interpretations

Suppose a scientist wanted to study the effects of viewing violent television on the aggressive behavior of children. The first step would be to _____ _____ _____ _____ . That might be: Does viewing violent television increase aggressive social behavior in nursery school children?

formulate a scientific
question

The next step is to _____ the variables and their relationships. That is, "watching violent television" might be defined to be having the children watch 30 minutes of Road Runner cartoons each day at 9 am for two weeks during nursery school. Similarly, "aggressive behavior" and all the other procedures of the research must be _____ .

operationalize

operationalized

Then comes the actual _____ _____ followed by the _____ _____ _____ _____ . As a result of these steps, the researcher will _____ _____ _____ _____ about the answer to the original question.

data collection
analysis of the data
draw
conclusions and interpretations

The skill in designing research is to create conditions so that the scientific question can be answered with some degree of certainty. For example, one requirement is that the variables of interest are measured accurately, that is, that those measurements have **reliability** and **validity**. When the measurement procedures assign the same value to a characteristic each time that it is measured under essentially the same circumstances, the measurement has _____ . If the procedures produce measurements that accurately reflect the conceptual variable being measured, they have _____ .

reliabilty

validity

For example, if two trained observers both rate the aggressiveness of children after observing them for several days and the ratings of the two observers correlate highly across the sample of children, then the measurement of aggressiveness has _____ between observers. If the ratings also correlate highly with the number of times each child verbally or physically attacks other children, then the ratings might be said to have _____ .

reliability

validity

A requirement of experimental research is that the study be designed so that it is reasonable to conclude that the independent variable caused or influenced the dependent variable. To do this, a simple experiment might consist of two groups, the **experimental** and the **control** groups. The treatment of interest is given to the _____ group. For example, the group of nursery school children shown Road Runner cartoons might be the _____ group. In contrast, another group of children would be given an experience as similar as possible to that of the experimental group except for the treatment of interest. It would be the _____ group.

experimental

experimental

control

It is necessary to define the control group in such a way as to minimize **confounding** by an **extraneous variable**. For example, if the control group watched no television in nursery school at all and if the experimental children were more aggressive than the control children, one would not know if they were more aggressive because they saw television—any television—or because they saw violent television in particular. In this case, the independent variable of violent television would be _____ with the _____ _____ of viewing television in general.

confounded; extraneous variable

Sometimes the extraneous variable resides in the subjects or in the observers. When the subjects of the research influence its outcome because they know, or think they know, something about the study, it is called **subject bias**. Sometimes, for example, parents know the researcher is studying parental behavior so they try to be "good" parents while being observed. They are introducing _____ _____ into the results. At other times, however, the **observers** might be the source of bias. For example, the raters might know that children who watched violent television "should" be more aggressive, so they might tend to rate them as being more aggressive. Then one has a case of _____ _____ .

subject bias

observer bias

To minimize these possibilities, research is sometimes designed to be **single** or **double blind**. When both subjects and observers do not know the research conditions, it is called a _____ _____ design. If one or the other, but not both, are ignorant about the research conditions, it is called a _____ _____ design.

double blind single blind

Finally, an important procedural method used in experimental studies is to **randomly assign** subjects to the experimental and control groups. Nursery school children, for example, differ in how aggressive they are before the scientific observations begin. If children are asked whether they want to watch Road Runner cartoons or Mister Rogers' Neighborhood, the aggressive children might all choose the cartoons. Any differences between the two groups would then be _____ with the fact that the children in the experimental group were more aggressive to begin with. To minimize this possibility, the researcher would _____ _____ children to the two groups.

confounded

randomly assign

     Throughout the design, conduct, and reporting of research, certain ethical principles must be followed. After the research is planned but before it is actually performed, a description of the proposed research must be reviewed by the institution's

_____ _____ _____ .

Internal Review Board

     The Board will decide issues of **risk/benefit, informed consent,** and **confidentiality**. For example, does the information to be gained from the proposed research outweigh any dangers to the participants? This is a question of _____ / _____ . The Board will also decide if the participants, who must read, understand, and sign a _____ or _____ form, will be given accurate and comprehensive information about the purpose, procedures, and possible risks and benefits of the research; that is, will the participants give truly _____ _____ ? Also, will the procedures protect the privacy of the participants and guarantee them _____ of information?

risk/benefit

consent; release

informed consent

confidentiality

     Researchers are also expected to report their procedures and results **honestly and forthrightly**. If an allegation is made that a researcher has not _____ and _____ reported the study, the institution's **scientific integrity procedures** may dictate a thorough review of the work. If the _____ _____ _____ result in a judgment of guilt, the consequence of committing _____ _____ can be severe, effectively ending a research or academic career.

honestly; forthrightly

scientific
integrity procedures
scientific misconduct

SELF-TEST

1. Methods that rely on direct systematic observations are called
   a. operational
   b. confounded
   c. empirical
   d. parametric

2. Research design consists of
   a. drawing a table of the groups to be studied
   b. the methods scientists use to make observations that will produce empirical information
   c. asking a scientific question
   d. analyzing the data with statistics

3. The "things" of scientific relationships are often called
   a. empirical concepts
   b. constants
   c. parameters
   d. variables

4. The variable whose values are thought to be influenced by another variable is called
   a. dependent
   b. independent
   c. parametric
   d. extraneous

5. If a researcher wants to describe a causal relationship, the research design should be
   a. observational
   b. nonparametric
   c. descriptive
   d. experimental

*6. A careful review of the research literature is likely to help in
   a. formulating the scientific question
   b. specifying parameters
   c. selecting control groups
   d. all of these

*7. A factor that influences a given behavior is a
   a. parameter
   b. extraneous variable
   c. independent variable
   d. all of these

*8. Factors in the research situation, but ones that are not related to the subjects, that influence a given behavior are sometimes called
   a. exogenous parameters
   b. endogenous parameters
   c. extraneous variables
   d. dependent variables

9. Specifying the actions required to measure a variable
   a. makes it an independent variable
   b. is its operational definition
   c. constitutes its parameters
   d. creates confounding

10. When the measurements accurately reflect the conceptual variable being measured, they are said to be
   a. reliable
   b. valid
   c. confounded
   d. parametric

11. When two observers tend to agree on the value of a variable for each member of a sample of subjects, the measurements are said to be
   a. endogenous
   b. exogenous
   c. reliable
   d. value

12. The group of subjects receiving the treatment of interest is called the
   a. experimental group
   b. endogenous group
   c. pilot group
   d. control group

13. A procedure that minimizes subject bias is
   a. random assignment of subjects to groups
   b. the double blind technique
   c. using a control group that consists of subjects selected in precisely the same way as those in the experimental group
   d. all of the above

14. The influence of an extraneous variable makes the design
   a. a single blind
   b. a double blind
   c. confounded
   d. experimental

*15. Placebos are sometimes used to control for
   a. exogenous confounds
   b. subject bias
   c. nonrandom assignment of subjects to groups
   d. sampling error

16. Generally, ethics in research is referred to as
   a. honesty and forthrightness
   b. informed consent
   c. scientific integrity
   d. confidentiality

17. Ethical procedures that typically transpire before the research is actually carried out are
   a. Internal Review Board approval
   b. obtaining informed consent

c. deciding risk/benefit   d. all of these

Questions preceded by an asterisk can be answered on the basis of the discussion in the text, but the discussion in this Study Guide does not answer them.

EXERCISES

1. Suppose a researcher was interested in determining the possible influence on children's aggressive social behavior of viewing violent television programs. The following experiment was designed. Two nursery schools agreed to participate. In Hillrise Nursery School, Mr. Gregory agreed to show Roadrunner cartoons for an hour each day beginning at 9 am to the 12 children ages 4–5 years in his class. At Greenbriar Nursery School, Mrs. Abernathy taught a class of 11 children ranging from 3 to 5 years of age. She said that she could not show cartoons to her children, but agreed to have the social behavior of her pupils observed.

    Cartoons were shown each weekday to Mr. Gregory's class for two weeks. On the last day, the cartoons were shown and then an observer came to watch the children play between 10 am and 11:30 am. The observer then ranked each child according to how aggressive he or she was during the observation period, with the rank of 1 going to the most aggressive child and the rank of

12 going to the least aggressive child. After lunch, the observer went to the Greenbriar school and observed the children there for an hour before their nap, and ranked those children in the same way as the children at Hillrise. The researcher then attempted to determine if watching violent television made the Hillrise children more aggressive than the Greenbriar children by comparing the average ranking for the two groups.

    Criticize this experiment. List and label all confounds and other problems that prevent the researcher from drawing a firm conclusion about the potential of watching violent television to produce aggressive social behavior in nursery school children.

2. Design a better experiment that would address the same research question posed above. Follow the steps of conducting a research project outlined in the text, and pay special attention to minimizing possible extraneous variables.

ANSWERS

Self-Test. (1) c, **(2)** b, **(3)** d, **(4)** a, **(5)** d, **(6)** d, **(7)** d, **(8)** a, **(9)** b, **(10)** b, **(11)** c, **(12)** a, **(13)** d, **(14)** c, **(15)** b, **(16)** c, **(17)** d.

Exercises. (1) The operational definition of watching violent television includes only watching cartoons and only Roadrunner cartoons. Therefore, the conclusion should be specific to Roadrunner cartoons, not all cartoons or violent television. Further, the operational definition of the control group involves not watching any kind

of television. Therefore, it is possible that watching any kind of television might influence social aggressiveness. A number of extraneous variables confound the possible results. For example, subject bias might be involved because neither the schools nor the children were randomly assigned to experimental and control groups. Notice also that the teachers volunteered to show cartoons or not show them. Further, observer bias could occur because the observer knew which group saw the cartoons. A number of exogenous extraneous variables were present as well. For example, experimental and control groups were confounded with specific nursery schools, classrooms, teachers, sex of teacher, time of day, and the time since eating or before a nap at which the assessment of aggressiveness was conducted. Finally, ranking the children within a class will not produce data that will reflect the relative amount of aggressiveness the children in the two groups displayed. The mean rankings will be different only because there was a different number of children in each group. **(2)** Discuss your proposed research design with your teacher or classmates.

CHAPTER 12

TOPICS IN PROBABILITY

CONCEPT GOALS

Be sure that you thoroughly understand the following concepts and how to use them in statistical applications.

- ◆ Set, element, subset, disjoint sets, union of sets, intersection of sets
- ◆ Classical probability, sample space, elementary event, mutually exclusive events
- ◆ Conditional probability, independent and dependent events
- ◆ Permutations, combinations

GUIDE TO MAJOR CONCEPTS

The cornerstone of inferential statistics is the concept of **probability**. A mathematical way of quantifying the likelihood of a given event is

_____ .   probability

Set Theory

The study of probability is made much easier by a knowledge of the terms of set theory. Objects or events can be thought of as being **elements** of **sets** or **subsets**.

A well-defined collection of things is called a _____ , and any   set
member of the collection is an _____ of that set. If we have   element
two sets, *A* and *B*, and every element of set *A* is also an element of set
B, then *A* is a _____ of *B*. A population is an example of a   subset
_____ , a sample from that population constitutes a _____ ,   set; subset
and a particular subject in the sample is an _____ .   element

Sets may have various relationships to one another. For example, they may be **equal** or **disjoint**. If every element of set *A* is also an element of set *B* and every element of *B* is also an element of *A*, then set *A* _____ set *B*; they are the same set. On the other hand, if   equals
no element of *A* is also in *B* and *no* element of *B* is also in *A*, then the sets are totally different from each other, and we say that they are

_____ .   disjoint

Sets may also share some, but perhaps not all, elements. For example, given two sets *A* and *B*, the set of all elements that are either in *A* or in *B* (and possibly in both *A* and *B*) is called the **union** of *A* and *B*, which is symbolized $A \cup B$. If set *A* includes all children aged 3 to 5 and set *B* includes all children aged 4 to 6, all children aged 3 to 6 constitutes the _____ of *A* and *B*, or in symbols,   union
_____ . In ordinary language, we frequently use the word *or* $A \cup B$
instead of *union*, as when we mention children age 3 to 5 _____ 4 to   or
6. In the space provided in Figure 12-1, draw two intersecting circles
A and *B* and shade the area representing the concept of union.

S

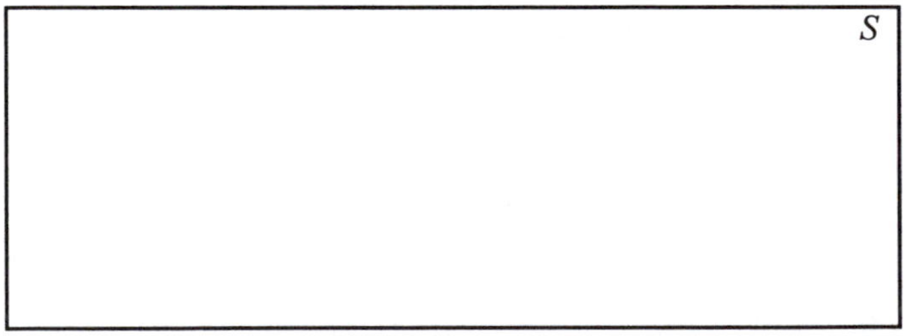

Fig. 12-1. Graphic representation of $A \cup B$. See Figure 12-3 in the text for the completed drawing.

Another important relationship is the **intersection** of two sets, symbolized by $A \cap B$. Given two sets, A and B, elements that are in both A and B (but not those in A or B alone) make up the _____ of A and B, which is written _____ . The intersection of the two sets of children age 3 to 5 and 4 to 6 would include children age _____ . In ordinary language we often use the word *and* rather than _____ , as when we designate that we want children who are in the group aged 3 to 5 _____ also in the group aged 4 to 6. In Figure 12-2, draw two intersecting circles A and B and shade the area representing the intersection of $A \cap B$.

> intersection; $A \cap B$
>
> 4 and 5
> intersection
> and

A special case arises when the intersection of two sets contains no elements at all , which set is called **empty** or the **null set** symbolized by $\varnothing$. When $A \cap B = $ __ , or the _____ set, A and B are said to be _____ . Also, if A is a set, then all elements not in A are said to be in the **complement** of A, written A' and said "not A." Therefore, $A \cap A' = $ _____ , or a set and its _____ are _____ . The set of all things being considered in any one discussion is the **universal set,** symbolized by S. Any set A within S is a subset of the _____ ____ ___ and $A \cup A'$ _____ .

> $\varnothing$; empty/null
> disjoint
>
> $\varnothing$
> complement; disjoint
>
> universal set S
> S

To review, let us start with people as the _____ _____ , and then let us divide the set *people* into two subsets, *men* and *women*.

> universal
> set

Mr. Jones is thus an _____ of the _____ of men, which in turn is a _____ of the set *people*. A group composed of either men (M) or women (W) is the _____ of those sets, which is symbolized by _____ . However, since no person is both a man and a woman, the _____ of the groups men and women, abbreviated _____ , has no elements, that is, it is the _____ set written _____ , and the sets are said to be _____ . Men is the _____ of the set women, and $M \cup W = $ _____ .

> element; set
> subset
> union
> $M \cup W$
> intersection
> $M \cap W$
> empty/null; $\varnothing$
> disjoint; complement
> S

Fig. 12-2. Graphic representation of $A \cap B$. See Figure 12-4 in the text for the completed drawing.

Classical Probability

In classical probability, all outcomes are equally likely to occur. Thus, the probability of obtaining a 3 in a single roll of a die is $\frac{1}{6}$ (*die* is the singular of *dice*) and the probability of a 6 is ___ .

$\frac{1}{6}$

    It will be helpful to formalize this commonplace example. The roll of a 3 is called an **elementary event**, and it is one of six possible elementary events that comprise the **sample space** of all possible outcomes of rolling a die. In the flip of a coin, obtaining a head is an _____ _____ in the _____ _____ consisting of heads and tails. Notice that an element of a set is now called an _____ _____ in a _____ _____ .

elementary event; sample space

elementary event; sample space

    When we wish to determine the probability that one of a set of elementary events (call it set A) within a sample space (label it set S) will occur, we need to determine how many such _____ _____ exist in A, symbolized by $\#(A)$, and how many elementary events exist in ____ , symbolized by $\#(S)$. The classical or simple probability of obtaining A is

elementary

events

S

$$P(A) = \frac{\#(A)}{\#(S)}$$

Therefore, the probability of A equals the number of elementary events in A, _____ , divided by the number of elementary events in the sample space S, _____ . Thus, the _____ of obtaining a head in a flip of a fair coin is ___ the probability of obtaining a 3 in a roll of a fair die is _____ , that of obtaining a 4 or less in a roll of a fair die _____ , that of drawing a spade from a standard deck of 52 cards is _____ , and that of dealing an ace as the first card from the deck is _____ . Notice that the numerical probability of any event must fall between _____ and ___ .

$\#(A)$

$\#(S)$; probability

$\frac{1}{2}$

$\frac{1}{6}$

$\frac{4}{6}$

$\frac{13}{52}$

$\frac{4}{52}$

0; 1

Conditional Probability

Conditional probability refers to situations in which the probability of an event is affected by the occurrence of another event. Suppose some children are to reach into a hat and blindly select one of ten numbers that correspond to each of ten different prizes the children will win. Johnny looks over the prizes and decides that he would most like to have a model racing car. If Johnny will draw first and we have no other information, all we can say is that Johnny has one chance in ten (or a probability of ___) of getting the racing car. But now suppose we know that he will be the second child to select a number. What is the probability that Johnny gets the racing car? The answer depends, or is **conditional**, upon what prize the first child

$\frac{1}{10}$

happens to pick. Thus, we may be interested in the probability that
Johnny gets the racing car *given* that the first child does not. This is a
_____ _____ . Given that the first child does
not get the car, the probability that Johnny does is ____. Given that
the first child indeed gets it, the probability that Johnny obtains it is
____.

    conditional probability

    $\frac{1}{9}$

    0

    The *probability* of Johnny's getting the car is said to be
conditional upon whether the first child does, but the two *events*
themselves are said to be **dependent events**. The events of drawing
an ace from a deck of cards on both the first and second draws when
cards are not replaced into the deck after drawing are two
_____ events, because the probability for the second
draw is _____ upon the outcome of the first draw. When
the probability of a particular event is *not* conditional on the outcome
of some other event, the two events are said to be **independent**. If
cards are replaced into the deck after a draw, then the probability of
an ace on the second draw does not depend on whether an ace was
drawn on the first. These two events are said to be _____
because the probability of the second is not _____ upon
the outcome of the first.

    dependent
    conditional

    independent
    conditional

Probability of $A \cap B$

Suppose we want to know the probability that two events will both
occur, such as drawing 2 kings in row without replacement from a
small deck of 12 cards containing the 4 kings, 4 queens, and 4 jacks.
The event of drawing two consecutive kings is a complex one,
consisting of the event A (obtaining a king on the first draw) *and*
event B (obtaining a king on the second draw given a king was drawn
on the first). The required probability is that both A *and* B occur,
which in set theory is called their _____ , symbolized by
_____ . The probability of the intersection of two events A and
B is

    intersection
    $A \cap B$

$$P(A \cap B) = P(A)P(B|A)$$

The probability that two events A and B both occur is the probability
of A *times* the conditional probability of B given that A has already
occurred. Applied to this example, the probability of drawing two
consecutive kings without replacement is the probability of obtaining
a king on the first draw (times/plus) _____ the probability of
obtaining a king on the second draw given that a king has already
been drawn. The probability of drawing the first king is $P(A) =$ ___ ;
the probability of drawing the second king given that a king was
obtained on the first draw, symbolized by _____ , equals ___ .

    times

    $\frac{4}{12}$

    $P(B|A)$; $\frac{3}{11}$

Therefore, the probability of both events (of the _____ of   intersection
A and B) equals

$$P(\text{_____})=\text{_____}$$
$$=\text{_____}=\text{_____}$$

$P(A\cap B)=P(A)P(B|A)$
$=\frac{4}{12}\cdot\frac{3}{11}=\frac{12}{132}=\frac{1}{11}$

Similarly, winning a four-team, single-elimination basketball
tournament (*single elimination* means that a team is eliminated when
it loses one game) requires a team to win two consecutive games.
Thus, event A is winning the first game; event B is winning the
second game given that you've won the first. The probability of
winning the first game is $P(A)=$ ____ . The probability of winning $\frac{1}{2}$
the second game given that you win the first (you don't get to play
the second without winning the first) is $P(B|A)=$ ____ . Thus, the $\frac{1}{2}$
probability of winning the tournament is symbolized by _____ $P(A\cap B)$
and equals _____ = _____ = _____ . $P(A)P(B|A)=\frac{1}{2}\cdot\frac{1}{2}=\frac{1}{4}$

When the probability of the intersection of two events is 0, the
events are said to be **mutually exclusive**. If M includes male
newborn infants and F includes female newborns, the two sets have
no common elements and are said to be **disjoint**. Since no baby is
born both male and female, the probability of $M\cap F$ is ___ , the sets   0
are _____ and the events of giving birth to a male and giving   disjoint
birth to a female are _____ _____ .   mutually exclusive

Probability of $A\cup B$

Suppose that we want to know the probability that either one *or* the
other of two events will occur. For example, what is the probability
of selecting either a queen or a red card from a deck composed of 12
face cards (i.e., the 4 jacks, queens, and kings)? Again this is a
complex event involving two simple events: event A, which is
_____ ____ _____ , and event B, which is _____   selecting a queen; selecting
____ _____ _____ . The desired probability is for either one or   a red card
the other (or both) events to occur, which in set theory is known as
the _____ of A and B, symbolized by _____ . The   union; $A\cup B$
probability is given by

$$P(A\cup B)=P(A)+P(B)-P(A\cap B)$$

The probability of A or B occurring is the probability of A *plus* the
probability of B *minus* the probability of A and B both occurring. In
the present problem, the probability of A is ___ and the probability of $\frac{1}{3}$
B is ___ . One more quantity is needed to solve the problem— the $\frac{1}{2}$
probability that both A and B occur, which is that of drawing a red
queen. There are two red queens in the deck of 12 cards, so
$P(A\cap B)=$ ____ . $\frac{2}{12}$

Thus the probability of selecting either a queen or a red card from the special deck, which can be found by the formula

_____ = _____ ,

is _____ = _____ = _____ .

$$P(A \cup B) = P(A) + P(B)$$
$$- P(A \cap B)$$

$$\tfrac{1}{3} + \tfrac{1}{2} - \tfrac{2}{12} = \tfrac{8}{12} = \tfrac{2}{3}$$

Similarly, the probability of getting an even number or a 1 in a single roll of a die equals the probability of the _____ of the events
A (which is _____ _____ _____ _____) and
B (which is _____ _____ _____). In this case there is no possibility of getting both an even number and a 1 (events A and B are _____ _____), so _____ $= 0$.
Therefore, the required probability is given by the formula
_____ = _____ , which equals

_____ = _____ = _____ .

union
obtaining an even number
obtaining a 1

mutually exclusive; $P(A \cap B)$

$$P(A \cup B) = P(A) + P(B)$$
$$- P(A \cap B)$$

$$\tfrac{1}{2} + \tfrac{1}{6} - 0 = \tfrac{4}{6} = \tfrac{2}{3}$$

Counting

To determine the probability of an event, it is necessary to count both the total number of events in the sample space and the number of events that would qualify as the desired outcome, i.e., to know the values of #S and #A in the formula $P(A) = \dfrac{\#A}{\#S}$. Sometimes this process can be tedious. An *ordered* set of objects or events is called a **permutation**, and the number of _____ of r objects that can be selected from a total of n objects is symbolized by $_nP_r$, which is read "the number of permutations of n things taken r at a time." Suppose there are ten dishes on a menu and you are interested in the number of different sequences of two dishes that you could have for dinner. Since you are concerned with the sequence in which you are to eat the dishes, you are interested in the number of _____ of ___ things taken ___ at a time, which can be symbolized _____ .

permutations

permutations
$10; 2;\ _{10}P_2$

The number of permutations of n things taken r at a time equals

$$_nP_r = \frac{n!}{(n-r)!}$$

Recall that $n!$, read "n factorial," means the product $n(n-1)(n-2)\ldots(1)$. In the present example, the number of permutations of ten things taken two at a time is

_____ = _____ = _____

= _____ = _____ .

$$_nP_r = \frac{n!}{(n-r)!} = \frac{10!}{(10-2)!}$$
$$= \frac{10 \cdot 9 \cdot 8!}{8!} = 90$$

Similarly, the number of sequences in which one could discard the first three cards of a seven-card rummy hand would be

$$\underline{\hspace{2cm}} = \underline{\hspace{3cm}} = \underline{\hspace{3cm}}$$

$$= \underline{\hspace{2cm}} = \underline{\hspace{2cm}}$$

$$_nP_r = \frac{n!}{(n-r)!} = \frac{7!}{(7-3)!}$$

$$= \frac{7 \cdot 6 \cdot 5 \cdot \cancel{4!}}{\cancel{4!}} = 210$$

However, suppose one is interested in the number of groups of r elements taken from a set of n, but that the order of elements within a group is not at all important. Such a group is called a **combination**, symbolized $_nC_r$. For instance, suppose a store has ten postcards depicting scenes from the local area, and you want to pick two of them. How many two-card combinations can be selected? The order in which you pick the two cards does not matter. The number of _____ of n things taken r at a time is

combinations

$$_nC_r = \frac{n!}{(n-r)!\,r!}$$

Applied to the problem of selecting two postcards from a set of ten, one has

$$\underline{\hspace{2cm}} = \underline{\hspace{3cm}} = \underline{\hspace{3cm}}$$

$$= \underline{\hspace{3cm}} = \underline{\hspace{3cm}}$$

$$_nC_r = \frac{n!}{(n-r)!\,r!} = \frac{10!}{(10-2)!2!}$$

$$= \frac{\overset{5}{\cancel{10}} \cdot 9 \cdot \cancel{8!}}{\cancel{8!} \cdot 2 \cdot 1} = 45$$

possible combinations. Similarly, the number of different committees of three people that could be selected from a group of five is given by

$$\underline{\hspace{2cm}} = \underline{\hspace{3cm}} = \underline{\hspace{3cm}}$$

$$= \underline{\hspace{3cm}} = \underline{\hspace{3cm}}$$

$$_nC_r = \frac{n!}{(n-r)!\,r!} = \frac{5!}{(5-3)!3!}$$

$$= \frac{5 \cdot \overset{2}{\cancel{4}} \cdot \cancel{3!}}{\cancel{2} \cdot 1 \cdot \cancel{3!}} = 10$$

To review, the difference between permutations and combinations is one of sequence or order. When the order of events is important, the number of _____ is required; but when order is unimportant, the number of _____ is needed. Thus, if the first three runners in a six-person preliminary heat of the hurdles will qualify for the final race, finding the number of different groups of finalists that are possible is a question involving _____ , but finding the number of win, place, and show (first, second, and third place) possibilities in a horse race is a problem involving _____ .

permutations
combinations

combinations

permutations

These methods of counting can be used to determine the probability of certain events. For example, what is the probability of randomly guessing which horse will finish first, which second, and which third in a six-horse race? First, because the order of finish is important, this is a problem involving _____ . Second, how many ways are there to pick the first three horses? Only one sequence of three horses will occur. Therefore, #(A) equals _____ . Third, the sample space consists of all possible permutations of ___ horses taken ___ at a time. Therefore, the #(S) is symbolized by ___ .

permutations

1

6

$3; \ _6P_3$

The required probability is

$$P(A) = \frac{\#(A)}{\#(S)} = \underline{\hspace{2cm}} = \underline{\hspace{2cm}}$$

$$= \underline{\hspace{1.5cm}} = \underline{\hspace{1.5cm}} = \underline{\hspace{1.5cm}}$$

$$\frac{1}{_6P_3} = \frac{1}{\frac{n!}{(n-r)!}}$$

$$\frac{1}{\frac{6!}{(6-3)!}} = \frac{1}{\frac{6 \cdot 5 \cdot 4 \cdot \cancel{3!}}{\cancel{3!}}} = \frac{1}{120}$$

What is the probability of being dealt all hearts in a five-card poker hand? The event A is receiving _____ _____ , but there is more than one way to accomplish that feat. Since the order of dealing is irrelevant, you need to determine the number of _____ of _____ hearts taken ____ at a time. Thus #(A) is symbolized by _____ and equals

all hearts

combinations; 13; 5

$_{13}C_5$

$$\underline{\hspace{2cm}} = \underline{\hspace{2cm}} = \underline{\hspace{2cm}}$$

$$= \underline{\hspace{2.5cm}}$$

$$= \underline{\hspace{1cm}} = \underline{\hspace{1.5cm}}$$

$$_{13}C_5 = \frac{n!}{(n-r)!r!} = \frac{13!}{(13-5)!5!}$$

$$= \frac{13 \cdot \cancel{12} \cdot 11 \cdot \cancel{10} \cdot 9 \cdot \cancel{8!}}{\cancel{8!} \cdot \cancel{5} \cdot \cancel{4} \cdot \cancel{3} \cdot \cancel{2} \cdot 1}$$

$$= 13 \cdot 11 \cdot 9 = 1287$$

However, the number of possible five-card hands constitutes the _____ _____ in this problem, and the actual number can be symbolized by _____ , which equals

sample space

$_{52}C_5$

$$\underline{\hspace{1.5cm}} = \underline{\hspace{1.5cm}} = \underline{\hspace{1.5cm}}$$

$$= \underline{\hspace{3cm}}$$

$$= \underline{\hspace{2cm}} = \underline{\hspace{1.5cm}}$$

$$_{52}C_5 = \frac{n!}{(n-r)!r!} = \frac{52!}{(52-5)!5!}$$

$$= \frac{52 \cdot 51 \cdot \overset{10}{\cancel{50}} \cdot 49 \cdot \overset{2}{\cancel{48}} \cdot \cancel{47!}}{\cancel{47!} \cdot \cancel{5} \cdot \cancel{4} \cdot \cancel{3} \cdot \cancel{2} \cdot 1}$$

$$= 52 \cdot 51 \cdot 10 \cdot 49 \cdot 2 = 2598960$$

Thus, the required probability in this case is

$$P(A) = \frac{\#(A)}{\#(S)} = \underline{\hspace{2cm}} = \underline{\hspace{3cm}} .$$

$$\frac{1287}{2598960} = .0005$$

A special application of determining the number of combinations occurs when there are only two possible outcomes — for example, success/failure, win/lose, rain/no rain — for each of several trials or occasions. This situation often calls for **binomial probability**. For example, flipping a coin five consecutive times is a case in which there are only two outcomes on a given flip, heads or tails, and a series of five trials or occasions, thus calling for _____ _____ .

binomial probability

Suppose at a carnival, you are given three tries to make a basket with a basketball. You must get at least two baskets to receive a prize, and your shooting percentage is 60%. What is the likelihood you win a prize? This probability is the union of getting two out of three and three out of three baskets. To calculate the probability of getting exactly two out of three, one needs to use _____

_____ .

binomial
probability

Specifically, in a sequence of n independent trials that have only two possible outcomes (success/failure) with the probability p of success and the probability q of failure $(q = 1 - p)$, the probability of exactly r successes in n trials is

$$P(r, n; p) = {}_nC_r p^r q^{n-r}$$

or

$$P(r, n; p) = \frac{n!}{r!(n-r)!} p^r q^{n-r}$$

So, to get exactly two out of three baskets, $n = $ ___ , $r = $ _____ , $p = $ ___ , and $q = $ ___ and the probability is given by

_____ = _____

= _____

= _____

3; 2
.60; .40

$P(2, \ 3; \ 60)$

$= \dfrac{3!}{2!(3-2)!}(.60)^2(.40)^{3-2}$

$= \dfrac{3 \cdot 2 \cdot 1}{2 \cdot 1 \cdot 1}(.60)^2(.40)^1$

$= .432$

Similarly, the probability of getting exactly three out of three baskets requires setting $n = $ ___ , $r = $ ___ , $p = $ ___ , and $q = $ ___ and solving (note that $0! = 1$ and $x^0 = 1$):

_____ = _____

= _____

= _____

3; 3; .60; .40

$P(3, \ 3; \ 60)$

$= \dfrac{3!}{3!(3-3)!}(.60)^3(.40)^{3-3}$

$= 1(.60)^3(.40)^0$

$= .216$

Therefore, the probability of getting at least two out of three is the union of two out of three (A) and three out of three (B) or

$$P(A \cup B) = P(A) + P(B)$$
$$= .432 + .216$$
$$= .648$$

Perhaps the game is worth a try — if you like the prize.

Table 12-1 presents a summary and guide for the probability concepts discussed in this chapter. Do these problems now, and follow the same format in working the exercises.

Table 12-1 Guided Computational Examples

1. **Probability**
   What is the probability of rolling a 4 or higher in one toss of a fair die?

   $A =$ _____ $\#(A) =$ _____

   $S =$ _____ $\#(S) =$ _____

   $P(A) = \dfrac{\#(A)}{\#(S)} =$ _____

2. **Probability of the Intersection of Two Events**
   What is the probability of drawing 2 aces in a row without replacement from a deck of 52 cards?

   $A =$ _____ $P(A) =$ _____

   $B|A =$ _____ $P(B|A) =$ _____

   $P(A \cap B) = P(A)P(B|A) =$ _____ $=$ _____

3. **Probability of the Union of Two Events**
   What is the probability of rolling an even number or obtaining a value greater than 4 with a single roll of a fair die?

   $A =$ _____ $P(A) =$ _____

   $B =$ _____ $P(B) =$ _____

   $A \cap B =$ _____ $P(A \cap B) =$ _____

   $P(A \cup B) = P(A) + P(B) - P(A \cap B) =$ _____ $=$ _____

4. **Permutations**
   How many seating arrangements are possible for seven people in a room containing four chairs?

   $n =$ _____ $r =$ _____

   $_nP_r = \dfrac{n!}{(n-r)!} =$ _____ $=$ _____

5. **Combinations**
   How many different groups from among nine people can sit in six chairs?

   $n =$ _____ $r =$ _____

   $_nC_r = \dfrac{n!}{(n-r)!r!} =$ _____ $=$ _____

6. **Binomial Probability**
   If the likelihood of rain is 40% on each of the next four days, what is the probability that it rains on two of the four days?

   $n =$ _____ $r =$ _____ $p =$ _____ $q =$ _____

   $P(r, n; p) = {}_nC_r p^r q^{n-r} = \dfrac{n!}{r!(n-r)!} p^r q^{n-r}$

   $=$ _____ $=$ _____ $=$ _____

SELF-TEST

1. In an introductory statistics class, 10 of the students, including Mary Smith, are psychology majors. Of these, 7 are also taking a course in learning. None of the students are taking experimental psychology. The three psychology majors who are not taking learning are studying clinical psychology. Two of the students taking clinical and three of the students taking learning are also enrolled in personality theory. Below are some set-theoretical concepts. Give an example of each from the situation described above.
   a.   the null set
   b.   disjoint sets
   c.   a set and a subset of it
   d.   the union of two sets
   e.   the intersection of two sets
   f.   the complement of a set
   g.   an element of a set

2. Define the simple probability of an event A.

3. What is the probability of the union of five mutually exclusive events if the complement of that union is the null set?

4. Are the following pairs of events mutually exclusive?
   a.   rolling an even number, and rolling less than a 5 with a fair die.
   b.   rolling an even number, and rolling less than 2 with a fair die.
   c.   drawing a heart and drawing a face card from a deck of 52 cards.

5. Which pairs of events are dependent and which are independent?
   a.   You select the chocolate cream pie from a cart containing one each of eight different desserts; you let your date pick a dessert first.
   b.   After drawing a king on your first draw, obtaining one ace in the next four draws without replacement.
   c.   Rolling a six after rolling four consecutive sixes with a fair die.
   d.   The baby is female in each of two consecutive births in a family.

*6. What is the relationship between conditional probability and independence?

7. Indicate whether each of the following illustrates the intersection or the union of two events.
   a.   rolling either 2 or 4 in a single roll of a die
   b.   drawing a card that is both red and even numbered
   c.   becoming a pilot in either the Army or the Air Force
   d.   answering each of 10 questions correctly

EXERCISES

1. In a special deck consisting of four aces, four kings, four queens, four jacks, and four jokers, determine the probability of drawing:
   a.   an ace
   b.   a red queen or a joker
   c.   a club (jokers are not considered to be members of any of the four suits)
   d.   the king of hearts followed by a joker with replacement after the first draw
   e.   the king of hearts followed by the king of spades without replacement
   f.   two kings in a row with replacement
   g.   two kings in a row without replacement

2. There are two baseball games being played this weekend. There is a forecast of 40% chance of rain for the Friday night game and a 30% chance for the Saturday contest.

   a. What is the probability that both games are rained out?

   b. What is the probability that they play both games?

  *c. What is the probability that one or no games are rained out?

  *d. What is the probability that exactly one game is rained out?

3. A teacher divides the class into two teams. Team *A* has three students, and team *B* has four. Each team member must work on a mathematics problem independently. A team wins if all members of the team complete the problem successfully. Suppose that the students in team *A* have probabilities of .65, .80, and .90 of solving the problem, whereas for team *B* these probabilities are .70, .75, .85, and .90.

   a. Which team would you bet on?

   b. What is the probability that neither team wins?

4. Since 10 of the applicants in the Campus Trivia-Talent Search Contest tie for the top score, a random drawing is conducted to determine who among the six men and four women will become the three people who will make up the team to be sent to the Collegiate Trivia Nationals.

   a. If the first person chosen for the team is a male, what is the probability that the second person will also be a male?

   b. If the first two team members are female, what is the probability that the third person drawn is a male?

   c. What is the probability that an all female team will be drawn?

5. Suppose Las Vegas bookies estimate the probability to be .20 that the horse named By-the-Nose will win the Triple Crown. The same bookies are estimating the probability of By-the-Nose winning both the first two races in the three-race Triple Crown sequence to be .50. Assuming that the three races are

independent events, what is the probability that By-the-Nose will win the third race?

6. Suppose that you apply for two summer jobs and that you know that you are one of a total of five applicants for the first and one of three for the second job. On a purely chance basis, what is the likelihood that you'll be unemployed this summer?

7. In a league of eight teams, how many different groups of teams could there be in a playoff tournament if the playoffs involved:

   a. four teams?

   b. two teams?

8. In joining a book club, you can choose any three books from a group of 12 titles. How many groups of books are possible?

9. How many orders of finish (first place through last) are possible in a six-horse race?

10. If there are 10 cars in a Grand Prix race, how many possible orders of finish for the first, second, and third place cars are there?

11. What is the probability of being dealt a five-card poker hand in which all cards are spades? In which all cards are of one suit? (No jokers are in the deck)

12. If a basketball coach has six guards, five forwards, and two centers, how many teams could be composed? (A basketball team has two guards, two forwards, and a center.)

13. Suppose that on a test of extra sensory perception a "receiving" subject is required to guess which of four geometric patterns a "transmitting" subject is viewing. The patterns are on four cards that are shuffled between each trial to obtain randomness. What is the probability that a receiver will be correct on at least three of four trials by chance alone?

14. Suppose a very long multiple-choice test contains questions with four alternative answers each, and your score will be the number right minus one-third the number wrong.

a. If you know absolutely nothing about the answer to a question, should you guess?
b. If you can eliminate one alternative as being definitely wrong, is it to your advantage to guess?

*Questions preceded by an asterisk can be answered on the basis of the discussion in the text, but the discussion in this Study Guide does not answer them.

ANSWERS

Table 12-1. (1) A = rolling a 4, 5, or 6; $\#(A) = 3$; S = rolling a 1, 2, 3, 4, 5, or 6; $\#(S) = 6$; $P(A) = \frac{3}{6} = \frac{1}{2}$. **(2)** A = an ace on the first draw; $P(A) = \frac{4}{52}$; $B|A$ = an ace on the second draw given that an ace was obtained on the first draw; $P(B|A) = \frac{3}{51}$; $P(A \cap B) = \left(\frac{4}{52}\right)\left(\frac{3}{51}\right) = \frac{1}{221}$. **(3)** A = rolling a 2, 4, or 6; $P(A) = \frac{3}{6}$; B = rolling a 5 or 6; $P(B) = \frac{2}{6}$; $A \cap B$ = rolling a 6; $P(A \cap B) = \frac{1}{6}$; $P(A \cup B) = \frac{3}{6} + \frac{2}{6} - \frac{1}{6} = \frac{2}{3}$. **(4)** $n = 7$; $r = 4$, $7!/(7-4)! = 840$. **(5)** $n = 9$; $r = 6$; $\dfrac{9!}{(9-6)!6!} = 84$. **(6)** $n = 4$, $r = 3$, $p = .40$, $q = .60$; $\dfrac{4!}{3!(4-3)!}(.40)^3(.60)^{4-3} = \dfrac{4 \cdot 3!}{3!1!}(.40)^3(.60) = .1536$

Self-Test. (1a) Students in the sample taking experimental psychology; **(1b)** students taking learning and students taking clinical; **(1c)** the two students enrolled in personality are a subset of the set of students taking clinical; **(1d)** students taking learning or clinical; **(1e)** students taking learning and personality; **(1f)** students not taking personality; **(1g)** Mary Smith. **(2)** $P(A) = \#(A)/\#(S)$. **(3)** 1.00. **(4a)** No; **(4b)** yes; **(4c)** no. **(5a)** Dependent; **(5b)** dependent; **(5c)** independent; **(5d)** independent. **(6)** A and B are independent if the conditional $P(B|A) = P(B)$. **(7a)** Union; **(7b)** intersection; **(7c)** union; **(7d)** intersection.

Exercises. (1a) $\frac{1}{5}$; **(1b)** $\frac{3}{10}$; **(1c)** $\frac{1}{5}$; **(1d)** $\frac{1}{100}$; **(1e)** $\frac{1}{380}$; **(1f)** $\frac{1}{25}$; **(1g)** $\frac{3}{95}$. **(2a)** $(.4)(.3) = .12$; **(2b)** $(1-.4)(1-.3) = .42$; **(2c)** $1-.12 = .88$ or $[P(\text{play both}) = (.6)(.7) = .42]$ plus $[P(\text{rain, play}) = (.4)(.7) = .28]$ plus $[P(\text{play, rain}) = (.6)(.3) = .18] = .88$; **(2d)** $.4(.7) + .6(.3) = .46$. **(3a)** $P(A) = .468$, $P(B) = .402$, bet on A; **(3b)** $P(A' \cap B') = (1-.468)(1-.402) = .32$. **(4a)** $\frac{5}{9}$; **(4b)** $\frac{6}{8}$; **(4c)** $\frac{4}{10} \cdot \frac{3}{9} \cdot \frac{2}{8} = \frac{1}{30}$. **(5)** $P(A)$ = win first two races = $.50$; B = win third race; $P(A \cup B) = P(A)P(B|A) = .20$; $P(B|A) = .40$. **(6)** $P(\text{unemployed}) = P(A')P(B') = \frac{4}{5} \cdot \frac{2}{6} = \frac{8}{15}$. **(7a)** $_8C_4 = 70$; $_8C_2 = 28$. **(8)** $_{12}C_3 = 220$. **(9)** $_6P_6 = 720$. **(10)** $_{10}P_3 = 720$. **(11)** $_{13}C_5/_{52}C_5 = \frac{33}{66640}$; $4\left(\frac{33}{66640}\right) = \frac{33}{16660}$. **(12)** $(_6C_2)(_5C_2)(_2C_1) = 300$. **(13)** $\dfrac{n!}{(n-r)!r!}p^r q^{n-r} = \dfrac{3}{64}$. **(14a)** For every four such questions on which you guess, you will get one correct and three wrong, yielding a score of $1 - \frac{1}{3}(3) = 0$, so there is no advantage or disadvantage in the long run to guessing; **(14b)** For every six such questions, you will get two correct and four wrong, yielding a score of $2 - \frac{1}{3}(4) = \frac{2}{3}$, so there is some advantage in the long run to guessing under these circumstances.

CHAPTER 13

TWO-FACTOR ANALYSIS OF VARIANCE

CONCEPT GOALS

Be sure that you thoroughly understand the following concepts and how to use them in statistical applications.

- ◆ Factor, level, cell
- ◆ Main effect, interaction
- ◆ Partitioning of variability
- ◆ Notation: X_{ijk}, N, n, p, q, T_{jk}, $T_{j.}$, $T_{.k}$, $T_{..}$, $\overline{X}_{j.}$, $\overline{X}_{.k}$, $\overline{X}_{..}$

GUIDE TO MAJOR CONCEPTS

In Chapter 10 you learned how to test the significance of the difference between a set of means for groups distinguished from one another by a single classification scheme. This is done by partitioning the total _____ in the sample into a component associated with differences _____ groups and a component associated with differences among individuals _____ groups. Both the between-groups and within-groups variance estimates, called _____ _____ , estimate the same population variance, assuming that _____ is true. Their ratio is distributed as the theoretical relative frequency distribution _____ , with degrees of freedom corresponding to the df associated with the MS in the _____ and in the _____ of the F ratio. Under H_0, F should vary only by sampling error, but if H_0 is wrong and the groups come from populations whose means differ from one another, F will probably be so (large/small) _____ that such a ratio is very unlikely to occur by _____ _____ alone. In that case, H_0 is _____ .

variability
between
within

mean squares
H_0
F

numerator; denominator

large
sampling error
rejected

Two-Factor Design

Suppose, however, that the subjects in the experiment can be simultaneously categorized according to two classification schemes, which in the analysis of variance are called **factors**. For example, suppose both male and female subjects are shown either an athletic, violent, or sensual film episode and the researcher is interested in the amount of the viewer's emotional involvement as measured by heart-rate levels. In this case, subjects can be classified as male or female, which defines the gender _____ . In addition, the particular film episode they view defines the film _____ . A subject is located in one and only one subgroup of each factor, called a **level** of the factor. Thus *male* and *female* are the _____ of the gender _____ ; the athletic, violent, and sensual film episodes constitute the _____ of the film _____ . Factors are usually abbreviated with capital letters, such as G and F for _____ and _____ (or A and B for unspecified factors), and the levels within a film factor are designated by the corresponding lower-case letter with subscripts (e.g., a_1, a_2). Thus, the two genders might be labeled ____ and ____ , while the three kinds of film would be labeled _____ , ____ , and ____ .

factor
factor

levels; factor

levels; factor
gender; film

g_1; g_2; f_1
f_2; f_3

What kinds of results might be obtained from such an experiment? The average heart rate for males and females might be different, or the heart rate for the three levels of the film factor might be significantly different from one another. If so, we say that there is a **main effect** for factor G or for factor F. Population mean differences between the levels of one factor, ignoring the levels of the other factor, constitute a _____ _____ . In a two-factor design such as this, one or both of the factors may exhibit a

_____ _____ .

main effect

main effect

However, suppose that the difference between levels of factor A is not the same within one level of factor B as it is within another level of factor B. This is called an **interaction**; that is, the nature of the effects of one factor interact with or depend on the levels of the other factor. Suppose males have higher heart rates than females during the athletic film but lower heart rates during the violent episode, while the sexes do not differ when watching sensual material. In this case there would be an _____ between factors G and F. In a two-factor design, then, there are three possible effects: two _____ _____ and one

_____ .

interaction

main effects
interaction

It is helpful to graph some of these possible results. In Figure 13-1 are six sets of axes. On each graph draw two lines, one for males and one for females, which reflect the mean heart rates for these groups under each film condition. Graph A is drawn for you, and it represents a _____ _____ for gender, no _____ _____ for film, and no _____ .

main effect
main effect; interaction

In the same manner, in graph B draw a result in which there is a significant main effect for film, no main effect for gender, and no interaction. In C, plot a situation in which both main effects are significant but there is no interaction; in D, a case in which both main effects and the interaction are significant; in E, a case in which only the interaction is significant; and in F, a case in which neither the main effects nor the interaction is significant.

Rationale

The rationale of two-factor analysis of variance is an extension of the logic of simple analysis of variance. Suppose that the score of a particular female watching the athletic film deviates 16 points from the grand mean over all subjects. What accounts for this? The deviation can be partitioned into four parts; the sum of the four parts equals the total deviation. First, suppose that the heart rate for females averages 10 points higher than the grand mean. One source of variability is thus the difference associated with being female, that

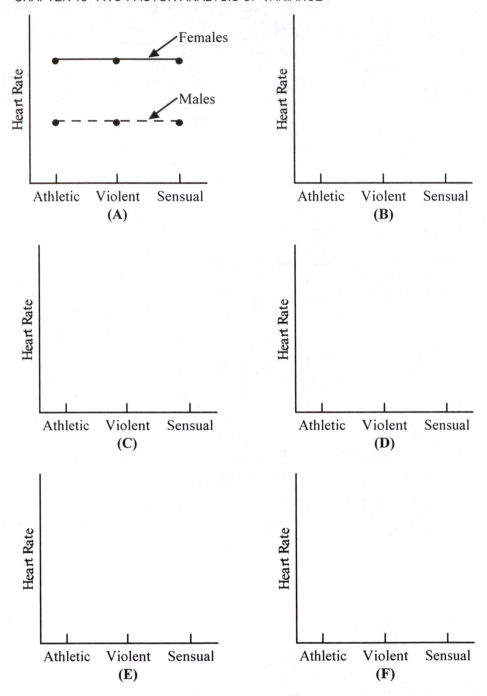

Fig. 13-1. Graphs of possible results of a two-factor study.

is, with _____ _____ . But this subject also watched the athletic film, and on the average subjects who saw the athletic film were 5 points lower than the grand mean; so a second source of variability is associated with the film condition or _____ _____ . Third, females watching the athletic events had heart rates that averaged 7 points above the grand mean even after general differences between the sexes and film conditions were considered. So a third source of variability is the _____ between G and F. Finally, 4 points separate this subject's score from the mean of her particular group. This variability about a group mean is known as _____ variability or error. The particular subject scored 16 points above the grand mean: +10 because she was a female, −5 because she watched the athletic film, +7 because of the unique interaction of the gender and film factors for that group and +4 because of individual differences and measurement error (within-group variation or error). Note that $16 = 10 - 5 + 7 + 4$.

factor G

factor F

interaction

within-group

When all these deviations are squared and summed over all subjects to produce **sums of squares**, we have that

$$SS_{total} = SS_{gender} + SS_{film} + SS_{gender \times film} + SS_{within}$$

The sum of squares of all deviations about the grand mean (SS_{total}) is equal to the sum of squares due to gender plus the sum of squares due to film plus the sum of squares due to the interaction between gender and film plus the sum of squares due to within-group variation. This is what is meant by **partitioning of variability**. In short, the _____ variability in the sample can be _____ into four sources: one part attributable to _____ _____ , another to _____ _____ , another to their _____ , and the remaining part to _____ variation. These sums of squares add to equal the total variability:

total
partitioned; factor G
factor F; interaction
within-group

$$\underline{\hspace{2cm}} = \underline{\hspace{2cm}} + \underline{\hspace{2cm}}$$

$$+ \underline{\hspace{2cm}} + \underline{\hspace{2cm}}$$

$SS_{total} = SS_{gender} + SS_{film}$
$+ SS_{gender \times film} + SS_{within}$

As in the case of simple analysis of variance, the null hypotheses for the main effects and the interaction are that these groups come from populations having the same means. Under these conditions, the four mean squares all estimate the same population variance. Restated in symbols, when _____ is true, _____ , _____ , and _____ all estimate the same population variance. But while MS_{gender}, MS_{film}, and $MS_{gender \times film}$ are all influenced by treatment effects, MS_{within} is not. Therefore, MS_{gender}, MS_{film}, and $MS_{gender \times film}$ can each serve as the numerator in an F ratio in which MS_{within} is the _____ . When compared to the F

H_0; MS_{gender}; MS_{film}
$MS_{gender \times film}$; MS_{within}

denominator

distribution, such ratiosallow us to determine the likelihood that the observed differences between means are a result of _____

_____ alone.

sampling
error

Notation

We will consider the notation used for a two-factor analysis of variance and then give a numerical illustration of these points. The mathematical notation for a score and the several formulas for an analysis of variance often appear frighteningly complex at first, but they are actually rather simple. Consider Table 13-1, which outlines the notation used in two-factor analysis of variance. Notice first that the data are arranged into rows and columns, with the levels of factor A constituting the rows and the levels of factor B the columns. A score is designated with three subscripts: X_{ijk} stands for the score of the ith subject in the jth level of factor A and the kth level of factor B.

Thus X_{312} would represent the score of the _____ person in the _____ level of A and the _____ level of B. Each group of scores is called a **cell** of the design, and ab_{23} is the notation for _____ 23 (read "cell two three"), which is the cell found at the intersection of the _____ row and _____ column. If there are p levels of factor A and q levels of factor B with n subjects per cell, the total number of cells in the design is _____ and the total number of subjects is _____ , or simply N. While the number of subjects in a cell is ____ . the total number of subjects in the entire analysis is ____ . Notice that the number of subjects is the same in each _____ .

The total of the scores in a single cell is symbolized by T_{jk}.

Thus, the total of the scores in cell 23 is labeled _____ , and T_{31} is the total of the scores in the group representing the _____ level of factor A and the ____ level of factor B. Similarly, the mean of the jkth cell is represented by $\overline{X}_{jk} = T_{jk}/n$, since there are n scores per

cell. The mean for cell 45 is written _____ = _____ . The total of

all scores in all cells in the jth row is $T_{j\cdot}$, in which the dot signifies that the scores have been summed over all columns. The same kind of notation is used for the mean of a row. Therefore, the total of all scores in row 3 is written ____ , so the mean of the scores in all cells in that row is symbolized by _____ . Since there are n subjects per group and q groups within a row, the formula for the mean of row 2

third
first; second

cell
second; third

pq
npq
n
N; cell

T_{23}

third
first

$\overline{X}_{45} = \dfrac{T_{45}}{n}$

$T_{3\cdot}$
$\overline{X}_{3\cdot}$

Table 13-1 Summary of Raw-Score Notation for the Two-Factor Analysis of Variance

	Factor B				
	b_1	b_2	$\cdots$	b_q	Row Means
a_1	X_{111} X_{211} X_{311} $\vdots$ X_{n11}	X_{112} X_{212} X_{312} $\vdots$ X_{n12}	$\cdots$ $\cdots$ $\cdots$ $\cdots$	X_{11q} X_{21q} X_{31q} $\vdots$ X_{n1q}	
	$\overline{X}_{11} = \frac{T_{11}}{n}$	$\overline{X}_{12} = \frac{T_{12}}{n}$	$\cdots$	$\overline{X}_{1q} = \frac{T_{1q}}{n}$	$\overline{X}_{1\cdot} = \frac{T_1}{nq}$
a_2	X_{121} X_{221} X_{321} $\vdots$ X_{n21}	X_{122} X_{222} X_{322} $\vdots$ X_{n22}	$\cdots$ $\cdots$ $\cdots$ $\cdots$	X_{12q} X_{22q} X_{32q} $\vdots$ X_{n2q}	
	$\overline{X}_{21} = \frac{T_{21}}{n}$	$\overline{X}_{22} = \frac{T_{22}}{n}$	$\cdots$	$\overline{X}_{2q} = \frac{T_{2q}}{n}$	$\overline{X}_{2\cdot} = \frac{T_2}{nq}$
$\vdots$	$\vdots$	$\vdots$	$\cdots$	$\vdots$	$\vdots$
a_p	X_{1p1} X_{2p1} X_{3p1} $\vdots$ X_{np1}	X_{1p2} X_{2p2} X_{3p2} $\vdots$ X_{np2}	$\cdots$ $\cdots$ $\cdots$ $\cdots$	X_{1pq} X_{2pq} X_{3pq} $\vdots$ X_{npq}	
	$\overline{X}_{p1} = \frac{T_{p1}}{n}$	$\overline{X}_{p2} = \frac{Tp_2}{n}$	$\cdots$	$\overline{X}_{pq} = \frac{T_{pq}}{n}$	$\overline{X}_{p\cdot} = \frac{T_{p\cdot}}{nq}$
Column Means	$\overline{X}_{\cdot1} = \frac{T_1}{n}$	$\overline{X}_{\cdot2} = \frac{T_2}{n}$	$\cdots$	$\overline{X}_{\cdot q} = \frac{T_q}{n}$	Grand Mean $\overline{X}.. = \frac{T..}{npq}$

Factor A

n = the number of subjects in each group
a_j = the jth level of factor A
b_k = the kth level of factor B
p = the number of levels of factor A
q = the number of levels of factor B

X_{ijk} = the ith score in the jkth group
T_{jk} = the total of all scores in the jkth group
$\overline{X}_{jk}$ = the mean of the jkth group
$T..$ = the total of all scores
$\overline{X}..$ = the grand mean

is _____ . By analogy, the total and mean of column 4 are $\qquad$ $\overline{X}_{2.} = \dfrac{T_{2.}}{nq}$

written respectively as _____ and _____ . Since the number of subjects $T_4 ; \overline{X}_4$

per group is _____ and there are p groups per column, the formula for n

the mean of column 1 is _____ . The grand total and grand $\overline{X}_1 = \dfrac{T_1}{np}$

mean of all subjects in the design are written with a subscript of two
dots, which indicate that the scores for subjects have been summed
over all rows and columns. The grand total is represented by _____ , $T..$
and the grand mean is _____ . $\overline{X}..$

Illustration

Now consider a numerical illustration of how the four mean squares
respond to different treatment effects. The data in Table 13-2 are
similar to a set of randomly selected numbers drawn from a single
population, except that they have been picked to make the
computations as easy as possible. Since they are random-like, all the
means estimate the same population mean and all the mean squares
estimate the same population variance. First, determine the means of
each cell, row, and column as well as the grand mean. Then calculate
SS_A, SS_B, SS_{AB}, and SS_{within} according to the formulas presented.
You should have found that $SS_A =$ _____ , $SS_B =$ _____ , 20;20
$SS_{AB} =$ _____ (very unusual), and $SS_{within} =$ _____ . 0; 100

   Now suppose that there is a population difference between the
levels of factor A. Table 13-3 presents the same data as in the
previous example, but 10 points have been added to each score in
row a_2 to reflect the population difference. Now recompute the
analysis of variance.

   As one might expect, the sum of squares for the factor A
increased over the first analysis. SS_A is now _____ , compared to 720
the previous value of _____ . Clearly, a population difference 20
between levels of A causes an (increase/decrease) _____ in increase
_____ . In contrast, the sums of squares for factor B and the AB SS_A
interaction (did/did not) _____ change. Observe also that did not
SS_{within} (is/is not) _____ the same as before. Thus, SS_{within} is
(is/is not) _____ sensitive to treatment effects. In short, a is not

Table 13-2 Numerical Illustration of Two-Factor Analysis of Variance with No Effects

Factor *B*

	b_1	b_2	
a_1	6 7 5 9 3 $\overline{X}_{11}=$	9 2 1 3 5 $\overline{X}_{12}=$	$\overline{X}_{1\cdot}=$
a_2	10 9 8 8 5 $\overline{X}_{21}=$	3 9 6 8 4 $\overline{X}_{22}=$	$\overline{X}_{2\cdot}=$
	$\overline{X}_{\cdot1}=$	$\overline{X}_{\cdot2}=$	$\overline{X}_{\cdot\cdot}=$

Factor *A*

$$SS_A = nq\sum_{j=1}^{p}\left(\overline{X}_j.-\overline{X}_{..}\right)^2$$

$$SS_A = (5)(2)\left[(5-6)^2 +(7-6)^2\right]$$

$$SS_A = \underline{\hspace{2cm}}$$

$$SS_B = np\sum_{k=1}^{q}\left(\overline{X}_{\cdot k} - \overline{X}_{..}\right)^2$$

$$SS_B = (5)(2)\left[(7-6)^2 +(5-6)^2\right]$$

$$SS_B = \underline{\hspace{2cm}}$$

Interaction

$$SS_{AB} = n\sum_{j=1}^{p}\sum_{k=1}^{q}\left(\overline{X}_{jk} - \overline{X}_j.-\overline{X}_{\cdot k}+\overline{X}..\right)^2$$

$$SS_{AB} = 5\left[(6-5-7+6)^2 +(4-5-5+6)^2\right.$$
$$\left.+(8-7-7+6)^2 +(6-7-5+6)^2\right]$$

$$SS_{AB} = \underline{\hspace{2cm}}$$

Within Groups

$$SS_{within} = \sum_{i=1}^{n}\sum_{j=1}^{p}\sum_{k=1}^{q}\left(X_{ijk} - \overline{X}_{jk}\right)^2$$

$$SS_{within} = \left[(6-6)^2 +(7-6)^2 +(5-6)^2+\cdots+(4-6)^2\right]$$

$$SS_{within} = \underline{\hspace{2cm}}$$

Table 13-3 Numerical Illustration of Two-Factor Analysis of Variance with an Effect for Factor A

Factor B

	b_1	b_2	
a_1	6 7 5 9 3 $\overline{X}_{11}=$	9 2 1 3 5 $\overline{X}_{12}=$	$\overline{X}_{1.}=$
a_2 (+10)	20 19 18 18 15 $\overline{X}_{21}=$	13 19 16 18 14 $\overline{X}_{22}=$	$\overline{X}_{2.}=$
	$\overline{X}_{.1}=$	$\overline{X}_{.2}=$	$\overline{X}_{..}=$

Factor A

$$SS_A = nq \sum_{j=1}^{p} \left(\overline{X}_{j.} - \overline{X}_{..} \right)^2$$

$$SS_A = (\ )(\ )\left[(\ \ )^2 + (\ \ )^2 \right]$$

$$SS_A = \underline{\qquad}$$

$$SS_B = np \sum_{k=1}^{q} \left(\overline{X}_{.k} - \overline{X}_{..} \right)^2$$

$$SS_B = (\ )(\ )\left[(\ \ \ )^2 + (\ \ \ )^2 \right]$$

$$SS_B = \underline{\qquad}$$

Interaction

$$SS_{AB} = n \sum_{j=1}^{p} \sum_{k=1}^{q} \left(\overline{X}_{jk} - \overline{X}_{j.} - \overline{X}_{.k} + \overline{X}_{..} \right)^2$$

$$SS_{AB} = (\ )\left[(\ \ )^2 + (\ \ )^2 + (\ \ )^2 + (\ \ )^2 \right]$$

$$SS_{AB} = \underline{\qquad}$$

Within Groups

$$SS_{within} = \sum_{i=1}^{n} \sum_{j=1}^{p} \sum_{k=1}^{q} \left(X_{ijk} - \overline{X}_{jk} \right)^2$$

$$SS_{within} = \underline{\qquad}$$

treatment effect in which the means for the levels of factor A are different from each other will influence only the size of the SS for _____ , whereas the SS for _____ , _____ , and _____ remain unchanged.

    Suppose an interaction exists in the population that can be represented in the data by adding 10 points only to cell ab_{21}. This has been done in Table 13-4. Complete this analysis of variance. Notice again that no charge occurred for the SS for _____ . However, not only did the interaction sum of squares rise from a former value of _____ to _____ , but SS_A and SS_B both (increased/decreased) _____ . The main effects changed, because adding 10 points to a single cell changes not only the mean of that cell but also the means of the respective row and column, whose values enter into the calculation of the _____ _____ . Because of this, when an analysis reveals a significant interaction, statisticians must be careful when interpreting a significant main effect.

factor A; B; AB

within

within

0; 125

increased

main effects

Computation

The computational procedures, which differ from the definitional formulas used above, are outlined in Table 13-5, which presents a guided computational example on another set of data. The data are presented in part A; determine the values of p, q, n, and N. Then in part B, form a table of totals, T_{jk}, which are the sums within cells, rows, and columns as well as the grand total. (You can also calculate a table of means with which you can interpret and graph the nature of any significant effects.) In part C, you must calculate five intermediate quantities. The first **(I)** is the grand total squared divided by the total number of scores in the sample, which is symbolized by _____ . The second quantity **(II)** is the sum of

$$\frac{T_{..}^2}{N}$$

all the squared scores, or _____ . The third **(III)** is

$$\sum_{i=1}^{n}\sum_{j=1}^{p}\sum_{k=1}^{q} X_{ijk}^2$$

found by summing the squared row totals and dividing the sum by nq, which is written _____ ; the fourth **(IV)** is found by

$$\sum_{j=1}^{p} T_{j.}^2 \big/ nq$$

Table 13-4 Numerical Illustration of a Two-Factor Analysis of Variance with an Interaction Effect

Factor *B*

Factor *A*	b_1	b_2	
a_1	6 7 5 9 3 $\overline{X}_{11}=$	9 2 1 3 5 $\overline{X}_{12}=$	$\overline{X}_{1\cdot}=$
a_2	(+10) 20 19 18 18 15 $\overline{X}_{21}=$	3 9 6 8 4 $\overline{X}_{22}=$	$\overline{X}_{2\cdot}=$
	$\overline{X}_{\cdot 1}=$	$\overline{X}_{\cdot 2}=$	$\overline{X}_{\cdot\cdot}=$

$$SS_A = nq\sum_{j=1}^{p}\left(\overline{X}_{j\cdot} - \overline{X}_{\cdot\cdot}\right)^2$$

$$SS_A = (\quad)(\quad)\left[(\qquad)^2 + (\qquad)^2\right]$$

$SS_A =$ __________

$$SS_B = np\sum_{k=1}^{q}\left(\overline{X}_{\cdot k} - \overline{X}_{\cdot\cdot}\right)^2$$

$$SS_B = (\quad)(\quad)\left[(\qquad)^2 + (\qquad)^2\right]$$

$SS_B =$ __________

Interaction

$$SS_{AB} = n\sum_{j=1}^{p}\sum_{k=1}^{q}\left(\overline{X}_{jk} - \overline{X}_{j\cdot} - \overline{X}_{\cdot k} + \overline{X}_{\cdot\cdot}\right)^2$$

$$SS_{AB} = (\quad)\left[(\quad)^2 + (\quad)^2 + (\quad)^2 + (\quad)^2\right]$$

$SS_{AB} =$ __________

Within Groups

$$SS_{\text{within}} = \sum_{i=1}^{n}\sum_{j=1}^{p}\sum_{k=1}^{q}\left(X_{ijk} - \overline{X}_{jk}\right)^2$$

$SS_{\text{within}} =$ __________

summing the squared column totals and dividing the sum by *np*, which is symbolized by __________ ; and the fifth (**V**) is found by summing the squared cell totals and dividing the sum by *n*, which can be expressed as __________ . The degrees of freedom and

$$\sum_{k=1}^{q} T_{\cdot k}^{2}/np$$

$$\sum_{j=1}^{p}\sum_{k=1}^{q} T_{jk}^{2}/n$$

Table 13-5 Guided Computational Example for Two-Factor Analysis of Variance

A. Data

Factor B

		b_1	b_2	b_3	b_4
		1	7	6	9
	a_1	4	10	9	7
		3	9	7	10
Factor A		1	8	8	8
		4	9	1	3
	a_2	1	10	0	4
		2	5	3	5
		3	8	2	3

$$p = \underline{\quad}, \qquad q = \underline{\quad}, \qquad n = \underline{\quad}, \qquad N = \underline{\quad},$$

B. Table of Totals

Factor B

Factor A		b_1	b_2	b_3	b_4	
	a_1	$T_{11} =$	$T_{12} =$	$T_{13} =$	$T_{14} =$	$T_1. =$
	a_2	$T_{21} =$	$T_{22} =$	$T_{23} =$	$T_{24} =$	$T_2. =$
		$T_{.1} =$	$T_{.2} =$	$T_{.3} =$	$T_{.4} =$	$T.. =$

C. Intermediate Quantities

$$\textbf{(I)} = \frac{T_{..}^2}{N} = \underline{\hspace{3cm}} \qquad\qquad \textbf{(II)} = \sum_{i=1}^{n}\sum_{j=1}^{p}\sum_{k=1}^{q} X_{ijk}^2 = \underline{\hspace{3cm}}$$

$$\textbf{(III)} = \frac{\sum_{j=1}^{p} T_{j.}^2}{nq} = \underline{\hspace{3cm}} \qquad\qquad \textbf{(IV)} = \frac{\sum_{k=1}^{q} T_{.k}^2}{np} = \underline{\hspace{3cm}}$$

$$\textbf{(V)} = \frac{\sum_{j=1}^{p}\sum_{k=1}^{q} T_{jk}^2}{n} = \underline{\hspace{3cm}}$$

D. Degrees of Freedom and Sums of Squares

$df_A = p - 1 = \underline{\hspace{3cm}}$ $SS_A = \textbf{(III)} - \textbf{(I)} = \underline{\hspace{3cm}}$

$df_B = q - 1 = \underline{\hspace{3cm}}$ $SS_B = \textbf{(IV)} - \textbf{(I)} = \underline{\hspace{3cm}}$

$df_{AB} = (p-1)(q-1) = \underline{\hspace{2cm}}$ $SS_{AB} = \textbf{(V)} + \textbf{(I)} - \textbf{(III)} - \textbf{(IV)} = \underline{\hspace{2cm}}$

$df_{within} = N - pq = \underline{\hspace{3cm}}$ $SS_{within} = \textbf{(II)} - \textbf{(V)} = \underline{\hspace{3cm}}$

$df_{total} = N - 1 = \underline{\hspace{3cm}}$ $SS_{total} = \textbf{(II)} - \textbf{(I)} = \underline{\hspace{3cm}}$

E. Summary Table

Source	df	SS	MS	F
A				
B				
AB				
Within				
Total				

For a significance level of .05 (from Table E, Appendix 2 of the text)

$A: df = p - 1$, $N - pq =$ _____ , _____ ; $F_{crit} =$ _____

$B: df = q - 1$, $N - pq =$ _____ , _____ ; $F_{crit} =$ _____

$AB: df = (p - 1)(q - 1)$, $N - pq =$ _____ , _____ ; $F_{crit} =$ _____

sums of squares are computed in part D, and the summary table is outlined in part E with each $MS = SS/df$ for that source and each F equal to the MS for that source divided by MS_{within}. The degrees of freedom are then used to determine the critical values from Table E in Appendix 2.

SELF-TEST

1. Define the terms *factor* and *level*, and relate them to the concept of main effect.

2. Name the three sources of variation in a two-factor analysis of variance that are associated with treatment effects.

3. Define *interaction*.

*4. In a two-factor analysis of variance, what implication does a significant interaction have for any significant main effects?

*5. What are the assumptions of the two-factor analysis of variance, and why is each necessary?

*6. State the three sets of hypotheses in a two-factor analysis of variance.

*Questions preceded by an asterisk can be answered on the basis of the discussion in the text, but the discussion in this Study Guide does not answer them.

EXERCISES

1. A physiological psychologist was studying the hormonal and genetic basis of gender differences in rough-and-tumble play among monkeys. Three groups of monkeys were observed for several days during their first year and another three groups during their

second year of life. The three groups were normal males, normal females, and animals that were genetically female but whose mothers had been treated with the male hormone testosterone while pregnant, thus producing a pseudomale. The time each animal spent in rough-and-tumble play is given below.[1] Evaluate these data.

and rearing circumstances on the smiling behavior of young babies observed infants in Israel of several different ages who were being reared either in a usual family environment, in a kibbutz, or in an orphanage. The number of smiles in an hour-long observation is given below for each infant in the study[2]. Evaluate the data.

	Factor *B* (Gender)		
	Males	Pseudomales	Females
First Year	10	9	2
	13	8	5
	14	8	4
Factor *A* (Age)	12	11	5
	11	7	4
Second Year	8	10	3
	14	7	6
	10	9	7

		Factor *B* (Age in Months)			
		2	4	8	16
Family		3	5	7	6
		4	9	6	7
		3	8	3	6
Factor *A* (Rearing Condition)	Kibbutz	4	6	7	3
		6	7	5	4
		2	5	3	5
Orphanage		6	7	5	4
		2	8	3	5
		3	5	6	4

2. A developmental psychologist who was interested in the effects of social stimulation

ANSWERS

Figure 13-1. See Figure 13-1 in your text. Your Graph B should be like the text's C, C like D, D like E, E like F, and F like A — but with three levels of film rather than two levels of vicarious reinforcement.

Table 13-5. (I) $= 903.125$, **(II)** $= 1218.000$, **(III)** $= 963.625$, **(IV)** $= 1051.750$, **(V)** $= 1169.500$; summary table, for factor A: $df = 1$, $SS = 60.500$, $MS = 60.500$, $F = 29.94**$; factor B: $df = 3$, $SS = 148.6250$, $MS = 49.5417$, $F = 24.52**$; interaction

$A \times B$: $df = 3$, $SS = 57.2500$, $MS = 19.0833$, $F = 9.44**$; within: $df = 24$, $SS = 48.5000$, $MS = 2.0208$; total: $df = 31$, $SS = 314.8750$; $F_{crit}(1, 24) = 4.26$, $F_{crit}(3, 24) = 3.01$.

Self-Test. (1) A factor is a basis of classification that includes two or more subgroups called levels. When there are significant differences in the population means for these levels, a main effect for this factor exists. **(2)** Main effect for factor A, main effect for factor B, $A \times B$ interaction effect. **(3)** An interaction exists when the nature of the effect for one factor is not the same under all levels of another factor. **(4)** A significant interaction indicates that any significant main effect probably should be qualified in its interpretation. **(5)** See discussion of *Assumptions of Two-Factor Analysis of Variance* in the text.

---

[1] Based on, but not identical to, a study by J. L Gewirtz, "The Cause of Infant Smiling in Four Child-Rearing Environments in Israel," in B. M. Foss, ed., *Determinants of Infant Behavior*, vol. 3 (London: Methuen, 1965).

[2] Based on, but not identical to, a study by R. W. Goy, "Organizing effects of androgen on the behavior of rhesus monkeys," in R. P. Michael, ed., *Endocrinology and Human Behavior* (Oxford: Oxford University Press, 1968).

(6) $H_0: \alpha_1 = \alpha_2 = \cdots = \alpha_j$, H_1: not H_0;

$H_0: \beta_1 = \beta_2 = \cdots = \beta_k$, H_2: not H_0;

$H_0: \alpha\beta_{11} = \alpha\beta_{12} = \alpha\beta_{21} = \cdots = \alpha\beta_{jk}$, H_1: not H_0.

Exercises. (1) (I) = 1617.0417, **(II)** = 1879, **(III)** = 1618.0833, **(IV)** = 1815.1250, **(V)** = 1822.7500; summary table, for age factor: $df = 1$, $SS = 1.0416$, $MS = 1.0416$, $F = .33$; gender factor: $df = 2$, $SS = 198.0833$, $MS = 99.0416$, $F = 31.6933**$; age $\times$ gender interaction: $df = 2$, $SS = 6.5834$, $MS = 3.2917$, $F = 1.0533$; within: $df = 18$, $SS = 56.2500$, $MS = 3.1250$; total: $df = 23$, $SS = 261.9583$. The amount of rough-and-tumble play seems to be related to testosterone level during and after the prenatal period. **(2) (I)** = 920.1111, **(II)** = 1032, **(III)** = 925.1667, **(IV)** = 961.1111, **(V)** = 974.6667; summary table, for rearing condition factor: $df = 2$, $SS = 5.0556$, $MS = 2.5278$, $F = 1.06$; age factor: $df = 3$, $SS = 41.0000$, $MS = 13.6667$, $F = 5.72**$; rearing $\times$ age interaction: $df = 6$, $SS = 8.5000$, $MS = 1.4167$, $F = .59$; within: $df = 24$, $SS = 57.3333$, $MS = 2.3889$; total: $df = 35$, $SS = 111.8889$. Age, but not rearing environment, was associated with smiling rates.

STATISTICAL PACKAGES

StataQuest
(To accompany the guided computational example in Table 13-5.)

StataQuest will compute a two-factor independent-groups analysis of variance for the sample data shown in Table 13-5. You must enter the dependent variable value and the levels of the *factors*, or independent variables, for that value on a separate line for each score in the dataset. In this example the dependent variable is Y, and the factors are A and B.

Start the StataQuest program. Bring up the Stata Editor window by clicking on the *Editor* button at the top of the screen.

Enter the value for the dependent measure Y in the first column of the spreadsheet, followed by the coded factor values for A and B in the second and third columns, respectively. There are two levels for factor A and four levels for factor B. Label the columns with the variable names Y, A, and B, respectively. To enter a variable name, point to the column heading and double click.

The data you have entered should appear on the spreadsheet as follows:

Y	A	B
1	1	1
4	1	1
3	1	1
1	1	1
7	1	2
10	1	2
9	1	2
8	1	2
6	1	3
9	1	3
7	1	3
8	1	3
9	1	4
7	1	4
10	1	4
8	1	4
4	2	1
1	2	1
2	2	1
3	2	1
9	2	2
10	2	2
5	2	2
8	2	2
1	2	3
0	2	3
3	2	3
2	2	3
3	2	4
4	2	4
5	2	4
3	2	4

To save the data that you entered:

> *File>Save as*
> > (Enter drive and file name for the save file.)
> > OK

Open a log file to save or print the results of the analysis. Click on the *Log* button at the top left of the screen, enter a name and drive, and click OK.

Carry out the two-factor analysis of variance, but do not select the plots:

> *Statistics>ANOVA>Two-way*
> > Dependent var.: Y   [click to transfer]
> > Cat. var. #1: A
> > Cat. var. #2: B
> > Include interaction: Yes
> > OK
> > Two-way ANOVA plots:
> > > Cancel

On the output for the analysis, the column headed "Partial SS" is simply the sum of squares for each source. Ignore the line labeled "model," and "Residual" means "within."

The mean, standard deviation, and n for each cell may be obtained as follows:

> *Summaries>Means and SDs by group>Two-way of means*
> > Data variable: Y   [click to transfer]
> > Row group variable: A
> > Column group variable: B
> > OK

Print the log file to obtain a copy of the output from the analyses:

> *File>Print Log*
> > OK

Exit StataQuest:

> *File>Exit*

StataQuest Program Output

```
. anova Y A B A*B

                    Number of obs =      32    R-squared     =   0.8460
                    Root MSE      = 1.42156    Adj R-squared =   0.8010

        Source |   Partial SS    df       MS             F      Prob > F
    -----------+----------------------------------------------------------
         Model |    266.375       7   38.0535714       18.83     0.0000
               |
             A |     60.50        1      60.50          29.94     0.0000
             B |    148.625       3   49.5416667        24.52     0.0000
           A*B |     57.25        3   19.0833333         9.44     0.0003
               |
      Residual |     48.50       24   2.02083333
    -----------+----------------------------------------------------------
         Total |    314.875      31   10.1572581


. tabulate A B, summ(Y)

            Means, Standard Deviations and Frequencies of Y

         | B
       A |        1           2           3           4       Total
    -----------+--------------------------------------------------+----------
       1 |      2.25         8.5         7.5         8.5  |    6.6875
         |       1.5    1.2909944   1.2909944   1.2909944  | 2.9375443
         |         4           4           4           4  |      16
    -----------+--------------------------------------------------+----------
       2 |       2.5           8         1.5        3.75  |    3.9375
         |  1.2909944   2.1602469   1.2909944   .95742711  | 2.8860296
         |         4           4           4           4  |      16
    -----------+--------------------------------------------------+----------
   Total |     2.375        8.25         4.5       6.125  |    5.3125
         | 1.3024702   1.6690459   3.4226139   2.7483761  | 3.1870454
         |         8           8           8           8  |      32
```

Minitab
(To accompany the guided computational example in Table 13-5.)

Minitab will perform a two-factor independent-groups analysis of variance using the General Linear Model command. The data must be presented in a "stacked" format. This approach requires entering the dependent variable value and the levels of the two factors for that value on a separate line for each score in the dataset. In this example, the dependent variable, or *response* variable, is Y, and the factors are A and B.

Start the Minitab program and click on the Data window. Using the data from Table 13-5, enter the values of the Y variable in the first column of the spreadsheet, followed by the levels of the two factors A and B in the second and third columns, respectively. Enter the variable names at the head of the three columns.

The data you have entered should appear on the spreadsheet as follows:

Y	A	B
1	1	1
4	1	1
3	1	1
1	1	1
7	1	2
10	1	2
9	1	2
8	1	2
6	1	3
9	1	3
7	1	3
8	1	3
9	1	4
7	1	4
10	1	4
8	1	4
4	2	1
1	2	1
2	2	1
3	2	1
9	2	2
10	2	2
5	2	2
8	2	2
1	2	3
0	2	3
3	2	3
2	2	3
3	2	4
4	2	4
5	2	4
3	2	4

To save the data that were entered:

> *File>Save Worksheet As*
>> (Enter drive and file name.)
>> OK

To carry out the two-factor analysis of variance, you must enter both factors and the interaction term exactly as shown:

> *Stat>ANOVA>General Linear Model*
>> Responses: Y   [double-click to transfer]
>> Model: **A B A*B**   [type in both factors and the interaction term]
>> OK

Then use the Cross Tabulation command to compute cell means:

> *Stat>Tables>Cross Tabulation*
>> Classification variables: A B   [double-click each to transfer]
>> Display: Counts
>> Summaries
>>> Associated variables: Y   [double-click to transfer]
>>> Display: Means, Standard deviations
>>> OK
>> OK

On the General Linear Model output, the ANOVA table for *Y* shows "Seq SS" and "Adj SS" which will be equal and simply the SS, "Adj MS" is the MS, and "Error" is Within. On the Tabulated Statistics output, the count *n*, mean, and standard deviation for each cell and for the marginals ("All") are shown.

Alternatively, the two-factor analysis of variance can be carried out using the Twoway command. To carry out the analysis using the Twoway command:

> *Stat>ANOVA>Twoway*
>> Response: Y   [double-click to transfer]
>> Row factor: A   Display means: Yes
>> Column factor: B Display means: Yes
>> OK

The Twoway command has the advantage that it plots means and 95% confidence intervals for the levels of each factor, and it is available on the student version of Minitab. Note, however, that its output does not give *F* ratios or associated *p* values for the main effect and interaction terms.

To obtain a copy of all the results, print the Session window:

> *File>Print Window*
>> Print Range: All
>> OK

Exit Minitab:

> *File>Exit*

Minitab Program Output

General Linear Model

```
Factor    Levels Values
A             2    1    2
B             4    1    2    3    4
```

Analysis of Variance for Y

Source	DF	Seq SS	Adj SS	Adj MS	F	P
A	1	60.500	60.500	60.500	29.94	0.000
B	3	148.625	148.625	49.542	24.52	0.000
A*B	3	57.250	57.250	19.083	9.44	0.000
Error	24	48.500	48.500	2.021		
Total	31	314.875				

Unusual Observations for Y

Obs	Y	Fit	StDev Fit	Residual	St Resid
23	5.0000	8.0000	0.7108	-3.0000	-2.44R

R denotes an observation with a large standardized residual

Tabulated Statistics

Rows: A   Columns: B

	1	2	3	4	All
1	4	4	4	4	16
	2.2500	8.5000	7.5000	8.5000	6.6875
	1.5000	1.2910	1.2910	1.2910	2.9375
2	4	4	4	4	16
	2.5000	8.0000	1.5000	3.7500	3.9375
	1.2910	2.1602	1.2910	0.9574	2.8860
All	8	8	8	8	32
	2.3750	8.2500	4.5000	6.1250	5.3125
	1.3025	1.6690	3.4226	2.7484	3.1870

```
Cell Contents --
              Count
          Y:Mean
            StDev
```

Two-way Analysis of Variance

```
Analysis of Variance for Y
Source          DF        SS          MS
A                1       60.50       60.50
B                3      148.63       49.54
Interaction      3       57.25       19.08
Error           24       48.50        2.02
Total           31      314.88
```

```
                            Individual 95% CI
A                 Mean    ----+---------+---------+---------+-------
1                 6.69                                (-----*-----)
2                 3.94       (-----*-----)
                            ----+---------+---------+---------+-------
                            3.60        4.80       6.00       7.20
```

```
                            Individual 95% CI
B                 Mean    ----+---------+---------+---------+-------
1                 2.38       (----*----)
2                 8.25                                (----*----)
3                 4.50                  (----*-----)
4                 6.12                       (-----*----)
                            ----+---------+---------+---------+-------
                            2.00        4.00       6.00       8.00
```

SPSS

(To accompany the guided computational example in Table 13-5.)

To compute a two-factor analysis of variance like the one shown in Table 13-5, you must first identify the dependent variable (the outcome measure) and the independent variables (factors). In this example the dependent variable is Y, and the factors are A and B.

Start the SPSS program. The Data Editor window should fill the screen.

Enter the dependent measure Y in the first column of the spreadsheet, followed by the coded factor values for A and B in the second and third columns, respectively. The levels for factor A run from 1 to 2 and for factor B from 1 to 4. Label the columns with the variable names Y, A, and B, respectively. To enter a variable name, point to the column heading and double click.

The data you have entered should appear on the spreadsheet as follows:

Y	A	B
1	1	1
4	1	1
3	1	1
1	1	1
7	1	2
10	1	2
9	1	2
8	1	2
6	1	3
9	1	3
7	1	3
8	1	3
9	1	4
7	1	4
10	1	4
8	1	4
4	2	1
1	2	1
2	2	1
3	2	1
9	2	2
10	2	2
5	2	2
8	2	2
1	2	3
0	2	3
3	2	3
2	2	3
3	2	4
4	2	4
5	2	4
3	2	4

To carry out the two-factor analysis of variance:

> *Statistics>General Linear Model>Simple Factorial*
>> Dependent: Y   [highlight and transfer]
>> Factor(s): A (? ?)
>> Define Range:
>>> minimum: **1**; maximum: **2**
>>> Continue
>> Factor(s): B (? ?)
>> Define Range:
>>> minimum: **1**; maximum: **4**
>>> Continue
>> OK

The output for this analysis gives more information than you need. The analysis of variance summary table given in the text is composed only of the headings and lines *A*, *B*, *A B*, Residual (that is, Within), and Total.

　　Means and standard deviations for each cell can be calculated as follows:

> *Statistics>Compare Means>Means*
>> Dependent List: Y   [highlight and transfer]
>> Independent List: A (Layer 1 of 2)
>> Next:
>> Independent List: B (Layer 2 of 2)
>> Options:
>>> Cell Statistics: Mean, Standard Deviation
>>> Continue
>> OK

The output is a table of means for the analysis.

　　To print the ANOVA table and the table of cell means:

> *File>Print*
>> Print range: All visible output.
>> OK

To save the data that you entered:

> *Window>SPSS Data Editor*
> *File>Save as*
>> (Enter device and file name for SPSS save file.)
>> Save

Exit SPSS:

> *File>Exit SPSS*

SPSS Program Output

Anova

Case Processing Summary[a]

	Cases					
	Included		Excluded		Total	
N	Percent	N	Percent	N	Percent
32	100.0%	0	.0%	32	100.0%

a. Y by A, B

ANOVA[a,b]

			Unique Method				
			Sum of Squares	df	Mean Square	F	Sig.
Y	Main Effects	(Combined)	209.125	4	52.281	25.871	.000
		A	60.500	1	60.500	29.938	.000
		B	148.625	3	49.542	24.515	.000
	2-Way Interactions	A * B	57.250	3	19.083	9.443	.000
	Model		266.375	7	38.054	18.831	.000
	Residual		48.500	24	2.021		
	Total		314.875	31	10.157		

a. Y by A, B

b. All effects entered simultaneously

Means

Case Processing Summary

	Cases					
	Included		Excluded		Total	
	N	Percent	N	Percent	N	Percent
Y * A * B	32	100.0%	0	.0%	32	100.0%

Report

Y			
1.00	1.00	Mean	2.2500
		Std. Deviation	1.5000
	2.00	Mean	8.5000
		Std. Deviation	1.2910
	3.00	Mean	7.5000
		Std. Deviation	1.2910
	4.00	Mean	8.5000
		Std. Deviation	1.2910
	Total	Mean	6.6875
		Std. Deviation	2.9375
2.00	1.00	Mean	2.5000
		Std. Deviation	1.2910
	2.00	Mean	8.0000
		Std. Deviation	2.1602
	3.00	Mean	1.5000
		Std. Deviation	1.2910
	4.00	Mean	3.7500
		Std. Deviation	.9574
	Total	Mean	3.9375
		Std. Deviation	2.8860
Total	1.00	Mean	2.3750
		Std. Deviation	1.3025
	2.00	Mean	8.2500
		Std. Deviation	1.6690
	3.00	Mean	4.5000
		Std. Deviation	3.4226
	4.00	Mean	6.1250
		Std. Deviation	2.7484
	Total	Mean	5.3125
		Std. Deviation	3.1870

CHAPTER 14

NONPARAMETRIC TECHNIQUES

CONCEPT GOALS

Be sure that you thoroughly understand the following concepts and how to use them in statistical applications.

- ◆ Parametric vs. nonparametric tests
- ◆ Pearson chi-square test
- ◆ Mann-Whitney U test for two independent samples
- ◆ Kruskal-Wallis test for k independent samples
- ◆ Wilcoxon test for two correlated samples
- ◆ Spearman rank-order correlation

GUIDE TO MAJOR CONCEPTS

The statistical techniques described in the previous chapters have
represented tests on parameters, such as μ or ρ. Such tests often
assume that the data consist of measurements made on at least an
interval scale, that the population distributions are normal, and that
homogeneity of variance prevails. Sometimes one cannot make such
assumptions about population parameters. Then it is helpful to have
some **nonparametric** statistical techniques available as alternatives
to the **parametric** ones discussed in previous chapters. The t test
between two means and the analysis of variance are _____   parametric
tests, whereas techniques that do not test the values of parameters
and that make different assumptions about the nature of the
populations involved are called _____ tests.   nonparametric

    It is not always clear when to use nonparametric rather than
parametric techniques. If the data consist of percentages of subjects
who did one thing or another or if subjects are ranked, then the data
are not measured with interval scales and one would probably choose
_____ methods. However, if the measurements are made   nonparametric
with interval scales but the distributions are skewed or variances
differ from group to group, the decision is not so clear. Generally, if
the data are only mildly deviant from the assumptions of _____   normal
distributions and _____ variances, one can still use   homogeneous
_____ tests; however, if departures from these assumptions are   parametric
severe, it may be best to use a _____ technique.   nonparametric

Pearson Chi-Square Test

Two decades ago, an opinion pollster wanted to know where people
placed the blame for the Watergate scandal. After President Nixon
held his first press conference following the beginning of the Senate
hearings in 1973, the pollster drew a sample and offered each person
five choices of where the blame should be put: (1) a plot against the
President, (2) the President himself, (3) usual dirty politics, (4) the
men around the President, or (5) no opinion. The result of such a
survey is the number (or percentage) of people questioned who
responded with each of the five alternatives. These data are on a
_____ scale of measurement and require a _____   nominal; nonparametric
analysis, in this case the **Pearson chi-square test**.

    The pollster suspects that opinion might depend or be
contingent upon whether the person answering the question is a
Republican, Democrat, or an Independent. For example, Democrats
may be more likely to blame the Republican President than other
groups. Statistically, one would like to determine the probability that
the distribution of responses over the five categories is the same for

each of the three samples of respondents (Republican, Democrats, Independents). That is, what is the probability that these distributions are not _____ upon group membership? The statistical question is addressed by the _____ _____ test.

      contingent
      Pearson chi-square

    Suppose that the results of this survey are those presented under "Computation" in Table 14-1. The row sums in the rightmost column and the column sums in the last row are called **marginals** (because they are in the margins of this table of data). They indicate how many Republicans, Democrats, and Independents were sampled and how many people, regardless of political affiliation, sided with each alternative.[1] Thus, the total sample included ____ people of whom Republicans numbered ____ , Democrats ____ , and Independents ____ . Regardless of political affiliation, ____ people blamed the President as indicated by the _____ of the table. The numbers within each cell of the table indicate the observed number of people with a particular affiliation who placed the blame on the circumstance indicated in that column. Thus, the President was blamed by ___ of the Democrats, while only ___ Republicans blamed him.

      1000
      350; 450
      200; 130
      marginals

      70; 20

    The results of the statistical technique can be used only to make inferences about a population whose characteristics are distributed in the proportions specified by the marginals in the table of sample data. Thus, the null hypothesis states: Given the observed _____ , the distribution of responses is identical for the three populations sampled (i.e., political affiliations). The alternative hypothesis states: _____ _____ _____ .

      marginals

      Given the observed marginals, these population distributions are not identical

    The statistical technique assumes that subjects were _____ and _____ sampled and that the groups of subjects are _____ of one another. Moreover, any single subject may qualify for **one and only one cell** or category. Thus, each person must have one political group affiliation and must choose only one of the five alternatives: Subjects must be in ____ and only one ____ . A final assumption of this test is that no expected frequency can be less than ___ , and in a 2 X 2 table, not less than ___ .

      randomly
      independently
      independent

      one; cell

      5; 10

---

[1] The distribution of response to these five categories regardless of political affiliation was reported by Daniel Yankelovitch, Inc., and published in *Time* (International Edition), 10 September 1973. The breakdown according to political affiliation is fictitious.

Table 14-1 Guided Computational Example for the $r \times c$ Chi-Square Test

Hypotheses

H_0: Given the observed _____ , the distributions of frequencies in the population
are not different for the groups.

H_1: Given the observed marginals, these _____ are different for these groups
(nondirectional/directional).

Assumptions and Conditions

1. The subjects are _____ and _____ sampled.
2. The groups are _____ .
3. Each observation qualifies for one and only one _____ .
4. No _____ frequency is less than _____ (if $r = c = 2$, not less than _____).

Decision Rules

Given a significance level of _____ , a _____ test, and $df = (r-1)(c-1) =$ _____ :

If _____ , _____ .

If _____ , _____ .

Computation

1. Place data into an $r \times c$ table.
2. Calculate the expected frequencies (if necessary) for each cell by taking the product of the
_____ and _____ marginal frequencies and dividing by _____ .

		Response Category				
	Plot	President	Politics	Appointees	No Opinion	Total
Republicans	$O = 60$	$O = 20$	$O = 130$	$O = 110$	$O = 30$	350
	$E =$	$E =$	$E =$	$E =$	$E =$	
Democrats	$O = 40$	$O = 70$	$O = 115$	$O = 200$	$O = 25$	450
	$E =$	$E =$	$E =$	$E =$	$E =$	
Independents	$O = 10$	$O = 40$	$O = 55$	$O = 60$	$O = 35$	200
	$E =$	$E =$	$E =$	$E =$	$E =$	
	110	130	300	370	90	1000

3. Check: Do expected frequencies add to equal the marginals?
4. Calculate χ^2_{obs}:

$$\chi^2_{obs} = \sum_{j=1}^{r} \sum_{k=1}^{c} \frac{\left(O_{jk} - E_{jk}\right)^2}{E_{jk}}$$

$$\chi^2_{obs} = \underline{\hspace{3cm}} = \underline{\hspace{3cm}}$$

where O_{jk} is the _____ frequency for the jkth cell

and E_{jk} is the _____ frequency for the jkth cell

Decision: _____ H_0 .

The decision rules are stated in terms of the statistic that will be calculated, χ^2, read "chi square." Critical values of the _____ _____ distribution are presented in Table G in Appendix 2 of your text. Look at that table now. The degrees of freedom are listed at the left, and critical values are given as a function of significance level and whether a directional or nondirectional test is appropriate. The degrees of freedom are given by $df = (r-1)(c-1)$ where r and c represent the number of rows and columns of the data table, respectively. For the polling example, $df = (\_\_\_)(\_\_\_)=\_\_\_$. The critical value for a nondirectional test at .05 is _____ . The chi-square distribution, like the F, is not symmetrical, and the computation is set up so that the null hypothesis is rejected if $\chi^2_{obs} \geq \chi^2_{crit}$. Therefore, the decision rules for this example are

If _____ , _____ .

If _____ , _____ .

chi-
square

$(3-1)(5-1) = 8$
15.51

$\chi_{obs}^{2} < 15.15$, do not reject H_0
$\chi_{obs}^{2} \geq 15.51$, reject H_0

The computation of χ^2_{obs} proceeds according to the outline in Table 14-1. First calculate the **expected** frequency for each score. The obtained or **observed** frequency for the jkth cell is symbolized by O_{jk}, while E_{jk} represents the _____ frequency. The E_{jk} for each cell is the product of the row and column marginals for that cell divided by the total number of subjects, N. Thus, for the first cell (Republicans who attributed the Watergate scandal to a plot against the President), the _____ frequency, symbolized for this cell by

expected

_____ , equals $(\_\_\_)(\_\_\_)/(\_\_\_)=\_\_\_$.

expected

E_{11}; $(350)(110)/(1000) = 38.5$

Similarly, the value of E_{23} equals $(\_\_\_)(\_\_\_)/(\_\_\_)=\_\_\_$. You can check the accuracy of your expected frequencies by determining whether they add up to the observed marginal totals.

$(450)(300)/(1000) = 135$

Once these expected frequencies are determined for each cell, χ^2_{obs} is given by

$$\chi^2_{obs} = \sum_{j=1}^{r}\sum_{k=1}^{c} \frac{\left(O_{jk} - E_{jk}\right)^2}{E_{jk}}$$

which directs one to take the difference between the _____ and _____ frequencies in a cell, _____ it, divide it by the _____ frequency for that cell, and _____ these quantities over all cells. For the first cell, $O_{11} = \_\_$, $E_{11} = \_\_$, $(O_{11} - E_{11}) = \_\_$, $(O_{11} - E_{11})^2 = \_\_$, and $(O_{11} - E_{11})^2/E_{11} = \_\_$; and so on for each cell. The total of such values over all cells equals $\chi^2_{obs} = \_\_$,

observed
expected; square
expected; sum
60; 38.5; 21.5
462.25; 12.0065
88.18

which conforms to the (first/second) _____ decision rule to _____ H_0. In this case, we conclude that the _____ of responses concerning Watergate are (the same/different) _____ for people having the three political affiliations. You might want to make a table of the percentage of each group responding to each of the five alternatives to see this more clearly. Now complete all parts of the guided computational example in Table 14-1.

second

reject; distributions

different

Mann-Whitney *U* Test for Two Independent Samples

Suppose that an experiment is conducted in which two independent groups of measurements are to be compared. Ordinarily, you would test the difference between the means of these groups with a _____ for _____ groups. But suppose that the measurements have been made on an ordinal scale or that the necessary assumptions regarding the population distributions cannot be met. In this case one might consider using a nonparametric analogue to the *t* test, such as the _____ test.

t test; independent

Mann-Whitney *U*

Consider the following experiment.[2] Suppose that some social scientists are interested in the effects of viewing different types of television programs on the behavior of nursery-school children. Two groups of children (from 3 1/2 to 5 1/2 years old) from low-income families are enrolled in a nursery school for six weeks. One group spends part of each school period watching such violent programs as *Batman* and *Superman*, while the other children view episodes from the program *Mister Rogers' Neighborhood*, which emphasizes sharing, cooperation, adaptive coping with frustration, and other forms of positive social behavior (which psychologists call *prosocial* behavior). The behavior of these children in the nursery-school setting is observed for a fixed period of time, and a panel of judges rates each child's prosocial behavior from 1 (minimum) to 10 (maximum). The data are presented under "Computation" in Table 14-2. The scores are presented in one row with each child's television group affiliation (*A* for violent or *B* for prosocial) shown above it. Notice that the scale of measurement is probably _____ in nature. Thus, a _____ statistical test, the _____ test, is selected.

ordinal
nonparametric
Mann-Whitney *U*

The rationale of the *U* test is simple. One ranks the scores, ignoring group affiliation. If there are no differences between groups, the average rank for each of the two groups should vary only by

---

[2] The data in this example are based upon, but not identical to, the results of a study conducted by L. K. Friedrich and A. H. Stein, "Aggressive and Prosocial Television Programs and the Natural Behavior of Preschool Children," *Monographs of the Society for Research in Child Development*, vol. 38, no. 151 (1973).

_____ _____ . If they differ more than could be   sampling error
expected by sampling error, then the null hypothesis (that the groups
are from identical _____) should be _____ .   populations; rejected

     The wording of the hypotheses for the U test is slightly different
than for the parametric t test. For the t test, the null hypothesis H_0 is
stated directly in terms of the population means: _____ . $\mu_1 = \mu_2$
However, in the U test, no assumptions are made about the
population variances, and consequently the null hypothesis must be
stated with respect to the entire form of the two **population
distributions**, not simply the means. Thus,

H_0: The population _____ from which the two samples are   distributions
     drawn are _____ .   identical

H_1: The _____ distributions are different in some way.   population

     The assumptions are that the subjects (or observations) are
_____ and _____ sampled and the two groups are   randomly; independently
_____ . In addition, it is necessary to assume that the variable   independent
underlying the scale of measurement is **continuous** and that the
measurement scale has at least **ordinal** properties. In the present
case, the ratings of prosocial behavior do form a scale with _____   ordinal
characteristics, and we can assume that they reflect a continuous
dimension of prosocial tendency in the children. Thus, the dependent
variable is _____ and the measurement scale is at   continuous
least _____ in nature.   ordinal

     The decision rules are stated somewhat differently, depending
upon whether the number of subjects in each group (n_A and n_B) is
more than 20. If n_A and $n_B \leq 20$, then you can determine the critical
values of U by consulting Table H in Appendix 2 of your text. Look
at this table now. Notice first that there is a different page for each
significance level. Therefore, you must locate the correct page
depending upon the chosen _____ _____ and   significance level
upon whether the test is directional or nondirectional. The rows and
columns represent the number of observations in each group. Simply
find the intersection of the row and column corresponding to n_A and
n_B. The two numbers located there are the upper and lower critical
values for U. Thus, in the present example, $n_A = $____ and $n_B = $____ ;   10; 8
for a nondirectional test at significance level .05, $U_{crit} = $ ____ and   17
____ . These critical values are similar to a confidence interval, so if   63
U_{obs} falls *between* these values, then H_0 is ____ _____ ; if U_{obs} not rejected
is *equal to or outside these limits*, then we _____ H_0. Formally,   reject

       If _____ $< U_{obs} <$ _____ , _____ .   17; 63; do not reject H_0
       If _____ or _____ , _____ . $U_{obs} \leq 17$; $U_{obs} \geq 63$; reject H_0

Table 14-2 Guided Computational Example for the Mann-Whitney U Test for Two Independent Samples

Hypotheses

    H_0: The population _____ from which the two samples are drawn are _____ .

    H_1: These _____ are different in some way (nondirectional/directional).

Assumptions and Conditions

    1.   The subjects are _____ and _____ sampled.

    2.   The two groups are _____ .

    3.   The variable is _____ and the measurement scale is at least _____ .

Decision Rules

If n_A or $n_B \leq 20$	If n_A or $n_B \geq 20$
Values of U_{obs} are found in Table H, Appendix 2.	U_{obs} is translated into a z_{obs} and the standard
Given $n_A =$ _____ , $n_B =$ _____ , a significance	normal distribution (Table A, Appendix 2) is used
level of _____ , and a _____ test:	with
If _____ $< U_{obs} <$ _____ , _____ H_0.	
If $U_{obs} \leq$ _____ or $U_{obs} \geq$ _____ ,	
_____ H_0.	

$$z_{obs} = \frac{U_{obs} - n_A n_B / 2}{\sqrt{\dfrac{(n_A)(n_B)(n_A + n_B + 1)}{12}}}$$

Computation (A = violent, B = prosocial)

Group	A	A	A	A	B	A	A	A	B	A	A	B	B	B	B	A	B	B
Score	1	2	2	4	5	5	5	6	6	6	7	8	9	9	9	9	10	10
Rank																		

Checks

$\left. \begin{array}{l} n_A = \underline{\hspace{1cm}} \\ n_B = \underline{\hspace{1cm}} \end{array} \right\} N = \underline{\hspace{1cm}}$
    $\left. \begin{array}{l} T_A = \underline{\hspace{1cm}} \\ T_B = \underline{\hspace{1cm}} \end{array} \right\} T_A + T_B = \underline{\hspace{1cm}}$

$$\text{Does } T_A + T_B \text{ equal } \frac{N(N+1)}{2} = \underline{\hspace{2cm}} \text{ ?}$$

$$U_{obs} = n_A n_B + \frac{n_A(n_A + 1)}{2} - T_A = \underline{\hspace{3cm}} = \underline{\hspace{2cm}}$$

Decision: _____ H_0.

If either n_A or n_B is greater than 20, Table H cannot be used. Instead, one would compute a z_{obs} with the formula given in Table 14-2 and use Table A in Appendix 2 of your text.

The computational procedures for U_{obs} (see Table 14-2) begin with placing the scores in order of increasing size regardless of group affiliation. Then they are ranked, the rank 1 being given to the lowest score. When scores are duplicated, all scores with the same value are given a rank equal to the average of the ranks these scores would have received if they had not been tied. Rank the scores in Table 14-2; when you get to the two scores of 2 you will see that these two scores, if not tied, would have received the ranks of ____ and ____ . Since they are tied, they are both assigned the average of these two ranks, which is ____ . But what rank is given to the next score, which is 4? Had the two scores of 2 not been tied, they would have occupied ranks 2 and 3, so the next assignable rank would be ____ . The score of 4 receives this rank. Rank the rest of the scores. You should have given the three scores of 6 the rank ____ and each of the scores of 9 the rank ____ .

    2; 3
    2.5

    4
    9
    14.5

You will need the values of n_A, n_B, N, T_A, and T_B. Calculate these in Table 14-2. The total number of subjects, N, equals the sum of the numbers in group A and group B, in this case ____ + ____ = ____ . T_A and T_B are the sum of the ranks for groups A and B. T_A = ____ and T_B = ____ .

    $10 + 8 = 18$
    65.5
    105.5

Many errors are made in ranking subjects in this and other parametric procedures. You can check your ranking accuracy with two tests. First, the highest score should receive, or share with a tied score, a rank equal to the total number of scores being ranked. In this case, the total number of scores is N = ____ . The two scores of 10 share ranks ____ and ____ and thus fulfill this check. Second, the total of all ranks of N scores should equal $N(N+1)/2$, in this case,

____ (____) / ____ = ____ , and $T_A + T_B$ should equal the same value: $T_A + T_B$ = ____ + ____ = ____ .

    18
    17; 18

    $18(18+1)/2 = 171$
    $65.5 + 105.5 = 171$

Now you are ready to calculate U_{obs}. Use the formula

$$U_{obs} = (n_A)(n_B) + \frac{n_A(n_A+1)}{2} - T_A$$

which for the present example equals

$$U_{obs} = \underline{\hspace{2cm}} = \underline{\hspace{2cm}} .$$

    $10(8) + \dfrac{10(11)}{2} - 65.5 = 69.5$

In this case, U_{obs} conforms to the (first/second) ____ decision rule, and H_0 is ____ . We conclude that the scores of children who saw violent television and of those who saw prosocial television are probably from population distributions that are different in some way. Now complete the guided computational example for this problem given in Table 14-2.

    second
    rejected

Kruskal-Wallis Test for k Independent Samples

Suppose that one cannot meet the assumptions of the simple analysis
of variance, perhaps because the distributions are not at all _____   normal
in form or the variances of the several groups are decidedly not
_____ . An alternative is the _____ test. In the   homogeneous; Kruskal-Wallis
experiment on television viewing and nursery-school behavior
described above, there were actually three groups of subjects
involved in the entire experiment: children who viewed the prosocial
"Mister Rogers' Neighborhood," those who saw the violent
programs, and those who saw neutral programs consisting of
travelogues and nature shows. The scientists observed not only
prosocial behavior but also aggressive actions among peers in the
nursery school. Does this television fare influence the amount of
verbal and physical aggression as expressed in a single aggression
index? The data are presented under "Computation" in Table 14-3.

 The nonparametric test to be illustrated is the _____   Kruskal-Wallis
test. The hypotheses and assumptions are similar to those for the
Mann-Whitney U test.

  H_0: The population _____ from which the groups are   distributions
   sampled are _____ .   identical
  H_0: The _____ distributions are different in some way.   population

 Again, notice that the hypotheses are stated in terms of
population _____ , not specifically in terms of their   distributions
_____ . It is necessary to assume that the subjects are   means
_____ and _____ sampled; that the k groups are   randomly; independently
_____ of one another; that the variable is _____ ;   independent; continuous
and that the scale of measurement has at least _____   ordinal
properties. The number of subjects in each group (n_j) must be at
least five (in symbols, _____) to have an accurate result. $n_j \geq 5$

 The statistic to be calculated is called H, but H is distributed as
χ^2 with $df = k - 1$ where k is the number of groups. For this
example, $df =$ _____ $=$ _____ . Given the .05 level of significance $3 - 1 = 2$
and a nondirectional test, the critical value of χ^2 (which is also H_{crit}
for this problem) given in Table G of Appendix 2 is $H_{\text{crit}} =$ ___ .   5.99
Therefore,

   If _____ , _____ . $H_{\text{obs}} < 5.99$, do not reject H_0
   If _____ , _____ . $H_{\text{obs}} \geq 5.99$, reject H_0

The computation of H_{obs} begins by ranking the scores without regard to group membership, assigning the rank 1 to the lowest score and treating ties as described above. Do this for the data in Table 14-3. It is most easily done by relisting in ascending order the scores in each group and then ranking across groups. Then determine the n_j and the total of the ranks (labeled _____) separately for each

T_j

each group. As before, check the accuracy of your ranking. Does the largest score (9) have a rank equal to N, the total number of subjects, in this case _____ ? And is the sum of the ranks for the three groups,

17

$T_1 + T_2 + T_3 =$ ____ + ____ + ____ = ____ equal to $N(N+1)/2$, which is _____ (_____)/_____ = _____ ?

47.5 + 54 + 51.5 = 153
17(17 + 1)/2 = 153

The value of H_{obs} is determined by obtaining the square of the total rankings of each group divided by its n_j and summing over all k groups. This value is symbolized by $\sum_{j=1}^{k}\left(\dfrac{T_j^{\,2}}{n_j}\right)$ and equals _____ .

1379.2917

N, the total number of subjects, equals _____ . These values may be substituted into the formula

17

$$H_{obs} = \left[\frac{12}{N(N+1)}\right]\left[\sum_{j=1}^{k}\frac{T_j^{\,2}}{n_j}\right] - 3(N+1)$$

$$= \underline{\hspace{5cm}}$$

$\dfrac{12}{[17(18)]}[1379.2917] - 3(18)$

which yields an H_{obs} of ____ . In this case, H_{obs} (is/is not) _____ large enough to reject H_0. We conclude that the observed differences in aggressive behavior between television groups could have resulted from _____ _____ . The different programs had no demonstrable effects on aggressive behavior.

.09
is not

sampling; error

Wilcoxon Test for Two Correlated Samples

Suppose that nine amateur springboard divers enroll in a special clinic designed to improve their form. The instructor rates them before and after the course. If we want to know whether the course has been effective in improving their diving form, we might use the parametric _____ for _____ samples. However, if we cannot meet the required assumptions, we might choose a nonparametric test such as the _____ test for two _____ samples. The data and format for this test are given in Table 14-4.

t test; correlated

Wilcoxon
correlated

Table 14-3 Guided Computational Example for the Kruskal-Wallis for k Independent Samples

---

Hypotheses

H_0: The population _____ from which the groups are sampled are _____ .

H_1: These _____ are different in some way (nondirectional/directional).

Assumptions and Conditions

1. The subjects are _____ and _____ sampled.
2. The k groups are _____ and all $n_j \geq$ _____ .
3. The variable is _____ and the measurement scale is at least _____ .

Decision Rules

Given a significance level of _____ , a _____ test, and the fact that H is distributed as χ^2 (Table G) with $df = k - 1 =$ _____ :

If _____ , _____

If _____ , _____

Computation

Prosocial		Neutral		Violent	
Score	Rank	Score	Rank	Score	Rank
7		9		3	
3		2		8	
8		4		8	
5		4		5	
6		7		6	
		7		1	
$n_1 =$	$T_1 =$	$n_2 =$	$T_2 =$	$n_3 =$	$T_3 =$
$\dfrac{T_1^2}{n_1} =$		$\dfrac{T_2^2}{n_2} =$		$\dfrac{T_3^2}{n_3} =$	

Check:

$$N = \sum_{j=1}^{k} n_j = \underline{\hspace{2cm}}, \quad \sum_{j=1}^{k} \frac{T_j^2}{n_j} = \underline{\hspace{2cm}}. \text{ Does } \sum T_j = \underline{\hspace{2cm}}$$

$$H_{\text{obs}} = \left[\frac{12}{N(N+1)} \right]\left[\sum_{j=1}^{k} \frac{T_j^2}{n_j} \right] - 3(N+1) = \underline{\hspace{2cm}} = \underline{\hspace{1cm}}. \text{ equal } \frac{N(N+1)}{2} = \underline{\hspace{1cm}} ?$$

Decision: _____ H_0.

The hypotheses for the Wilcoxon test are similar to those of previous tests.

H_0: The population _____ for the two correlated   distributions

groups of observations are _____ .   identical

H_1: The _____ distributions are _____ in some way.   population; different

It is necessary to assume that the pairs of observations are _____ and _____ sampled; a pair consists of   randomly; independently

observations made either on the _____ subject or on closely   same

_____ subjects. In this example, a pair consists of one diver's   matched

scores before and after instruction. Finally, the scale of measurement must be at least _____ in nature; this applies not only to the   ordinal

scores within a pair, but also to the differences between scores in a pair.

The test statistic for the _____ test is W. If N is 50 or   Wilcoxon

less, then the critical value of W is given in Table I in Appendix 2 of your text. Look at this table now. It gives the value of W_{crit} for various N, for different significance levels, and for directional and nondirectional tests. In contrast to other tables, if W_{obs} is *less than or equal to* the tabled value of W_{crit}, then the null hypothesis is rejected. Thus, for $N = 9$ and a directional test (training programs rarely lead to poorer performance) at the .05 level, the critical value of W_{crit}

would be ___ . If W_{obs} is larger than this value, _____   8; do not reject

H_0; if W_{obs} is less than or equal to this value, _____ H_0. However,   reject

if N is greater than 50, Table I in your text cannot be used. In this case, a z_{obs} can be calculated (see Table 14-4) and the critical values obtained from Table A in Appendix 2.

The computation of W_{obs} follows the outline under "Computation" in Table 14-4. First, compute d_i, the difference in value between the two scores in each pair. In the next column, write the absolute value of the d_i. Then rank these $|d_i|$, giving the smallest $|d_i|$ rank 1; finally, in the last column, affix the sign of the original d_i

to this ranking. For the first subject, $d_i =$___ , $|d_i| =$___ , the rank in   −.7; .7

the total sample will be 7, and since the original d_i was negative, the

signed rank will be ___ . However, ranking for the Wilcoxon test is   −7

slightly different from ranking for other tests. If the difference between a pair of measurements is zero, then that pair is eliminated, is not given a ranking, and is not counted in the total N, which is the original number of pairs minus the number of pairs for which $d_i = 0$. Complete the ranking of the scores in Table 14-4 now. The total of

Table 14-4 Guided Computational Example for the Wilcoxon Test for Two Correlated Samples

Hypotheses

H_0: The population _____ for the two correlated groups of observations are _____ .

H_1: These _____ are different in some way (nondirectional/directional).

Assumptions

1. The pairs of observations are _____ and _____ sampled, but the two observations of a pair are made on the _____ or _____ subjects.

2. The measurement scale is _____ both for individual scores and for d_i, the differences between the scores in each pair.

Decision Rules

If $N < 50$	If $N > 50$

The critical values of W_{obs} are found in Table I.

Given $N =$ _____ , a significance level of

_____ , and a _____ test:

If _____ > _____ , _____ .

If _____ ≤ _____ , _____ .

W_{obs} is translated into a z_{obs}, and the standard normal distribution (Table A) is used with

$$z_{obs} = \frac{W_{obs} - N(N+1)/4}{\sqrt{\dfrac{N(N+1)(2N+1)}{24}}}$$

Computation

Before	After	d_i	$|d_i|$	Rank of $|d_i|$	Signed Rank of $|d_i|$
4.5	5.2				
3.0	4.5				
6.2	5.9				
4.9	5.2				
5.5	5.8				
5.8	6.1				
6.1	7.2				
7.0	7.1				
5.0	5.6				

$N =$ original number of pairs minus
    the number of pairs for which $d_i = 0$

$N =$ _____

$T_+ =$ _____
$T_- =$ _____

Check: Does $T_+ + T_- =$ _____ equal $\dfrac{N(N+1)}{2}$?

$W_{obs} =$ _____

Decision: _____ H_0 .

the positive ranks is symbolized by T_+ and equals ____ ; the total of the negative rankings (without the minus signs) is called T_- and equals ____ . Again, you should check the accuracy of your rankings by determining that $T_+ + T_-$ equals $N(N+1)/2$. In this case, these two quantities should both be ____ .

3.5

41.5

45

If there is no difference between the two populations of measurements, T_+ and T_- should be nearly equal. But if the populations differ from one another, then there will be a difference between T_+ and T_-. As this difference becomes greater, one of the T's will grow large and the other quite small. W_{obs} is the smaller of T_+ and T_-; in this case, $W_{obs} =$ _____ . To reject H_0 and conclude that the clinic produced improvement, this value must be _____ than or _____ to W_{crit}; otherwise, do not reject H_0. Since W_{crit} is 8, the decision is to _____ H_0; that is, the clinic has helped. Complete Table 14-4 now.

3.5

less;equal
reject

Spearman Rank-Order Correlation

It is sometimes preferable to use a nonparametric measure of correlation rather than the Pearson product-moment correlation coefficient presented in Chapter 6. Such a nonparametric index is the _____ _____ correlation coefficient, symbolized by r_S. Perhaps the distributions of scores are not normal or the measurements are so crude that you trust only the ordinal properties of the scale. In these cases, you might want to use the _____ _____ correlation coefficient, or ____ .

Spearman rank-order

Spearman rank-order; r_S

In the study of television programming and aggressive behavior described previously, one might be interested in whether there is a relationship between the aggression of the child when he or she comes to the nursery school and the amount of increase in aggressive behavior which occurs after seeing the violent TV programs. Suppose trained raters rank order 10 nursery-school children on general aggressiveness before the experiment begins. Then, the number of aggressive acts committed by each child is counted before and after the two-week period in which violent television programs are shown in the school. Table 14-5 presents the general aggressiveness ranking (1 = least aggressive) and the amount of increase in aggressive behavior over the two-week period.

To compute r_S you must rank the scores in each column separately. In this case, the scores under "general aggression" are already ranked, so these rankings need only be repeated under "rank aggression." However, the scores under "increment in aggression" must be ranked. The ranking is done in the same manner as for the

Mann-Whitney U test, and ties are given average ranks. Complete this ranking in Table 14-5 now. In the next columns, the differences in the ranks for each pair of measures (d_i) is calculated, and then this value is squared (d_i^2) in the final column. Sum these squared

differences in ranks to obtain $\sum_{i=1}^{N} d_i^2$, which equals ___ . N is the   43.5

number of pairs *including* those having no difference in their rankings; in this case, $N =$ ___ . The value of r_S is given by the   10
formula

$$r_S = 1 - \left[\frac{6 \sum_{i=1}^{N} d_i^2}{N^3 - N} \right]$$

$$r_S = \underline{\hspace{2cm}} = \underline{\hspace{2cm}}.$$ $1 - \left[\dfrac{6(43.5)}{1000 - 10} \right] = 1 - .26$

which yields a value of $r_S =$ ___ . This rank-order correlation   .74
coefficient can be interpreted in the same manner as the Pearson r.
Thus, r_S can take on values between ___ and ___ , the absence of $-1.00; +1.00$
any linear relationship between the measures is indicated by $r_S =$ ___ ,   0
and the proportion of variability shared by the two variables is given
by ___ . r_S^2

   To test whether the observed value of r_S is really from a
population in which $\rho_S = .00$, one need only to consult Table J in
Appendix 2 of the text, which gives the critical values of r_S for
specified N, various significance levels, and directional and
nondirectional tests. For this example, the critical value of r_S with
$N = 10$ for a nondirectional test at significance level .05 is ___ .   .649
Therefore, we _____ H_0. Complete Table 14-5 at this time.   reject

Table 14-5 Guided Computational Example for the Spearman Rank-Order Correlation Coefficient

Hypotheses

H_0: _____

H_1: _____ (nondirectional/directional)

Assumptions and Conditions

1. The subjects are _____ and _____ sampled.
2. The scales of measurement have at least _____ properties.

Decision Rules

Given a significance level of _____ , $N =$ _____ pairs of scores, and a _____ test
(see Table J, Appendix 2 of the text):

If _____ , _____ .

If _____ , _____ .

Computation

Subject	General Aggression Rank	Increment in Aggression	Rank of Aggression	Rank of Increment	d_i	d_i^2
A	6	−1				
B	2	−3				
C	3	8				
D	8	10				
E	9	17				
F	1	7				
G	4	5				
H	7	16				
I	10	14				
J	5	8				

N = number of pairs of measurements including zero differences = _____ $\sum_{i=1}^{N} d_i^2 =$ _____

$$r_S = 1 - \left[\frac{6 \sum_{i=1}^{N} d_i^2}{N^3 - N} \right] = \underline{\hspace{2cm}} = \underline{\hspace{2cm}}$$

$r_S =$ _____

Decision: _____ H_0 .

SELF-TEST

1. Under what circumstances might a nonparametric test be preferred over a parametric test?

*2. If a sample is relatively small and contains one very extreme score, why might you prefer to use a nonparametric test?

3. In general, how do null hypotheses in parametric tests of the differences between groups differ from those in nonparametric tests?

4. Which nonparametric statistical test would be appropriate in the following situations?

   a. A sampling of 25 graduate students and 24 professors participated in a blind taste test, trying to distinguish between a local American beer and an expensive European import after a skeptical student asked why the professor always paid twice as much for imported beer when the difference in taste is not detectable anyway. Are students and professors different in their ability to identify the two beers? (One professor participated in the experiment but claimed that the results were irrelevant, since people never drink beer blindfolded.)

   b. To examine the relationship between creativity and need for affiliation, the art projects of 30 eighth-grade students were ranked on creativity by a panel of artists. The same students were given a Need-for-Affiliation test which contained items such as "I like to spend several hours each day by myself."

   c. A group of 25 15-day-old rats are given a diet that contains a small amount of lead. Another group of 25 rats are given the same diet without the lead. Each infant rat in the lead group is matched to a litter-mate of the same sex and weight in the non-lead group. As adults, each animal is tested for number of errors in a visual discrimination task.

   d. To test the common belief that boys initiate aggressive behavior more often than girls do, 16 boys and 14 girls were observed in a nursery school for a week. Each child's aggressive instigations were counted, and the groups were compared.

   e. It is predicted that left-handed students are more likely to major in a "quantitative" as opposed to "verbal" area than right-handed students. One hundred left-handed and 100 right-handed students are polled as to their majors. The major areas of study are divided into "strongly quantitative," "strongly verbal," and "mixed."

   f. People are said to differ in social assertiveness as a function of their general body build. Three groups of 10 men were selected according to their general physique (endomorph, mesomorph, ectomorph) and then given a test of social assertiveness to assess this notion.

5. Given a nondirectional alternative, the .05 level of significance, and the tables in Appendix 2 of the text, specify the critical values and decision rules for each of the situations in question 4 above. (Check your answers to question 4 first, and assume that there are no rank differences of 0.)

*6. The Mann-Whitney U test is sometimes said to be the nonparametric version of the t test. However, the two statistical tests are different in some ways. Discuss the important differences.

7. In what test are subjects eliminated from the analysis if their data do not help to discriminate between groups?

*Questions preceded by an asterisk can be answered on the basis of the discussion in the text, but the discussion in the Study Guide does not answer them.

EXERCISES

1. When scores originally measured on interval or ratio scales are converted to ranks, some information is lost. When the distributions of the raw scores are skewed or have different variability, these characteristics are modified by the ranking process; therefore, a parametric test and nonparametric test on the same data may yield different results. Below is a small set of data from two independent groups of subjects. Perform both a parametric *t* test and a Mann-Whitney *U* test on these data. Note how the extreme scores and skewness of the original measurements are altered when the data are ranked. Contrast the results of the two statistical tests, compare this difference with the difference between the mean and median of a distribution, and discuss the importance of this characteristic for the choice a researcher sometimes must make between parametric and nonparametric tests.

Original Data		Ranking of Data	
A	B	A	B
1	4		
2	6		
2	6		
3	9		
3	10		
5	12		
5	12		
8	14		
15	46		

2. Ethnologists report that there seems to be a nearly universal tendency for women to smile more frequently than men. To examine this tendency among American teenagers, the photographs in a junior high school yearbook were classified as smiling (with teeth showing) or not smiling (no teeth showing).[3] Are the differences shown in the following data significant?

	Smiling	Not Smiling
Boys	118	28
Girls	112	5

3. A social scientist was concerned about the accuracy of feedback that students in urban schools receive from teachers concerning their academic performance. Two high schools were selected, one in which the students have a good record of academic performance and one in which they do not. Twelve juniors are randomly selected from the good school and 14 from the poor school. They are given a standardized test of academic achievement and are then asked to rate on a 10-point scale how well they are doing in school (10 is very good). The results are given below.[4] Using nonparametric techniques first test the hypothesis that there is no difference in test performance in the two samples. Then determine if there is no difference in the students' perceptions of how well they are doing in school. What is the relationship between test performance and judgment of school performance for each of these two groups? Are these values different from zero? What do these results seem to be saying about the feedback these students are receiving about their school performance?

---

[3] Data based on the 1985 yearbook of Valley View Junior High, Omaha, Nebraska.

[4] Based on, but not identical to, research reported by Sanford M. Dornbush, "Racism Without Racists: Institutional Racism in Urban Schools." *The Black Scholar* 7 (November 1975); also appears in Stanford Center for Research and Development in Teaching, occasional paper no. 8 (November 1975).

Good School		Poor School	
Achievement Test	Performance Rating	Achievement Test	Performance Rating
92	2	87	7
97	3	99	10
84	6	103	10
122	7	84	7
113	5	90	6
99	4	78	4
102	5	81	6
128	6	95	9
114	7	76	5
109	5	85	5
117	4	95	7
		93	8
		78	5

4. In a special summer tutoring program, teenagers who are from low-income families and who do not have summer jobs are assigned to tutor grade-school youngsters three times per week. Since it was hoped that the program would provide benefits to the tutors as well as to the youngsters, the class rankings of the tutors on a standardized academic achievement test for the school year prior to the tutoring experience were compared to the class rankings in the school year after the experience. Suppose the following data were obtained. Did the tutors increase their own academic performance relative to their peers?

Class Rankings		
Tutor	Before	After
A	34	29
B	54	54
C	82	86
D	17	9
E	23	19
F	36	12
G	66	57
H	7	5
I	119	95

5. For more than 60 years the Fels Research Institute has conducted a study of children from birth through adulthood. As part of this study, IQ tests were given periodically throughout each child's life, and the child's parents were rated on a variety of dimensions. Children could be classified as showing an increase, no change, or a decrease in IQ from ages 3 to 17. Below are the ratings made on the children's parents with respect to how much they attempted to accelerate the mental development of their children.[5] Determine whether there is any association between pattern of IQ change and parental attempts at acceleration.

Decline in IQ	No Change in IQ	Increase in IQ
75	81	91
78	86	95
86	89	89
79	85	99
85	88	98
78	89	86
87	80	84

---

[5] Inspired by R. B. McCall, M. I. Appelbaum, and P. S. Hogarty, "Developmental Changes in Mental Performance," *Monographs of the Society for Research in Child Development* vol. 38, no. 150 (1973).

*6. At a casino, a given die showed over 120 rolls the following pattern of scores from 1 to 6: 17, 21, 28, 15, 19, 20. Test whether the die is fair.

ANSWERS

Tables 14-1 to 14-5. See corresponding tables in text.

Self-Test. (1) When the assumptions made by parametric tests cannot be met, a nonparametric alternative may be appropriate. Typically this occurs when the scale of measurement is nominal or ordinal or when the assumptions of normality or homogeneity of variance cannot be met. **(2)** Nonparametric tests usually use only the ordinal characteristics of the scale of measurement and would therefore minimize the influence of an extreme score on the result. The extreme score is also likely to make the distributions deviate from normality and the variances not be homogeneous. **(3)** In a parametric test, the nature of the null hypothesis is: population value = population value. In a nonparametric test the null states: population distribution = population distribution. **(4a)** 2×2 chi square contingency table; **(4b)** Spearman rank-order correlation; **(4c)** Wilcoxon; **(4d)** Mann-Whitney U; **(4e)** 2×3 chi square contingency table; **(4f)** Kruskal-Wallis. **(5a)** If $\chi^2_{obs} < 3.84$, do not reject H_0; if $\chi^2_{obs} \geq 3.84$, reject H_0; **(5b)** If $r_S < .363$, do not reject H_0; if $r_S \geq .363$, reject H_0; **(5c)** If $W_{obs} > 89$, do not reject H_0; if $W_{obs} \leq 89$, reject H_0; **(5d)** If $64 < U_{obs} < 160$, do not reject H_0; if $U_{obs} \leq 64$ or $U_{obs} \geq 160$, reject H_0; **(5e)** If $\chi^2_{obs} < 5.99$, do not reject H_0; if $\chi^2_{obs} \geq 5.99$, reject H_0; **(5f)** If $H_{obs} < 5.99$, do not reject H_0, if $H_{obs} \geq 5.99$, reject H_0. **(6)** The t test examines the difference between means directly, uses measurement information higher than the ordinal level, and has somewhat more power-efficiency; the U test compares all aspects of the two distributions, considers only ordinal information from the measurement scale, and has good power-efficiency, although somewhat less than t. **(7)** The Wilcoxon test.

Exercises. (1) $t_{obs} = 1.86$ (or 1.87 with rounding), t_{crit} (significance level $= .05$, $df = 16$) $= 2.120$, do not reject H_0. $U_{obs} = 68$, U_{obs} (significance level $= .05$, $n_1 = n_2 = 9$) $= 17, 64$; reject H_0. The ranking reduces the influence of an extreme score (46) on the estimate of the variance of the sampling distribution (the denominator in the t formula) and reduces skewness and heterogeneity of variance in this example. Thus, when there are a few extreme scores, nonparametric tests may be more sensitive to group differences than parametric tests. **(2)** $\chi^2_{obs} = 13.15$, $\chi^2_{crit} = 3.84$ for .05 level, $df = 1$; reject H_0. **(3)** For achievement test performance: $U_{obs} = 18$, $U_{crit} = 37, 106$ for .05 level, nondirectional, $n = 11, 13$; reject H_0. For performance ratings: $U_{obs} = 109.5$, $U_{crit} = 37,106$ for .05 level, nondirectional, $n = 11, 13$ reject H_0. For the good school: $r_S = .53$; critical value $= .619$ for .05 level, nondirectional, $N = 11$; do not reject H_0. For the poor school: $r_S = .88$, critical value $= .561$ for .05 level, nondirectional, $N = 13$; reject H_0. It appears that students in poor schools are actually doing more poorly but they think they are doing well. Maybe their teachers are telling them they are doing well, and/or maybe they think they are doing well because they are doing well enough relative to other students in their school. Their performance ratings do correlate well with their achievement scores. **(4)** $W_{obs} = 2.5$, $W_{crit} = 3$, .05 level, nondirectional, $N = 8$; reject H_0. **(5)** $H_{obs} = 8.96$ (rounding errors can be large, maintain at least three significant digits), $H_{crit} = 5.99$, .05 level, nondirectional, $df = 2$; reject H_0. **(6)** $\chi^2_{obs} = 5.00$; $\chi^2_{crit} = 11.07$, .05 level, nondirectional, $df = 4$; do not reject H_0, the evidence is not sufficient to question the fairness of the die.

STATISTICAL PACKAGES

StataQuest

Chi-Square Test (Table 14-1)

StataQuest does not have a menu command to compute chi-square for a row-by-column set of frequency data, such as in Table 14-1. The analysis can be accomplished, however, by entering a command in the Stata command window. Data entry requires three input variables: one for the row identification, one for the column identification, and one that contains the frequencies for each cell.

Start the StataQuest program. Bring up the Stata Editor window by clicking on the *Editor* button at the top of the screen. Enter the frequency data from Table 14-1 for each of the 15 cells, one cell at a time. The cell frequency, *Freq*, goes in the first variable or column on the spreadsheet. The row number, *Row*, goes in the second variable, and the column number, *Col*, goes in the third variable. Label the columns with the variable names *Freq*, *Row*, and *Col*. The data you have entered should appear on the spreadsheet as follows:

Freq	Row	Col
60	1	1
20	1	2
130	1	3
110	1	4
30	1	5
40	2	1
70	2	2
115	2	3
200	2	4
25	2	5
10	3	1
40	3	2
55	3	3
60	3	4
35	3	5

To save the data that you entered:

> *File>Save as*
> > (Enter drive and file name for the save file.)
> > OK

Open a log file to save or print the results of the analysis. Click on the *Log* button at the top left of the screen, enter a name and drive, and click OK. Only one log file is needed for all five exercises in this chapter.

The Stata command *tabulate* will set up the contingency table, and the *chi2* keyword will calculate the chi-square statistic to test for independence between the table variables. The *fweight=Freq* command function will weight the cases by cell frequency, so that in effect it is reading information for all 1000 cases (the total number of persons sampled) rather than the 15 cells that are entered on the spreadsheet.

To carry out the analysis, enter the following command in the Stata Command window at the bottom of the StataQuest screen *exactly* as it appears below:

      tabulate Row Col [fweight=Freq], chi2

Row, Col, and *Freq* are the names of the three variables that you entered into the spreadsheet. Hit *Enter* to register the command.

The output from this command lists the frequency in each cell followed by the chi-square statistic and its associated probability value. Expected frequencies are not listed by this command.

Clear the spreadsheet to input data for the next example:

     File>New
       OK

StataQuest Program Output

```
. tabulate Row Col [fweight=Freq], chi2

       | Col
  Row|         1            2            3            4            5 |    Total
  -----+----------------------------------------------------------------+--------
    1 |        60           20          130          110           30 |      350
    2 |        40           70          115          200           25 |      450
    3 |        10           40           55           60           35 |      200
  -----+----------------------------------------------------------------+--------
 Total|       110          130          300          370           90 |     1000

 Pearson chi2(8) =    88.1818     Pr = 0.000
```

Mann-Whitney *U* Test (Table 14-2)

StataQuest uses the Mann-Whitney *U* Test to compare two independent samples. In the case of the example in Table 14-2, observers rated children's positive behavior after watching one of two different television programs, violent or prosocial. The dependent or test variable is the behavior rating, *Behavior*; the coded treatment or grouping variable is the program type, *Program* (1=Prosocial; 2=Violent).

Bring up the Stata Editor window by clicking on the *Editor* button at the top of the screen. Enter the values for the behavior ratings in the first column and the code for the type of programming in the second column. Label the two columns with the variable names *Behavior* and *Program*. The data in the spreadsheet should now appear as follows:

Behavior	Program
1	1
2	1
2	1
4	1
5	1
5	1
6	1
6	1
7	1
9	1
5	2
6	2
8	2
9	2
9	2
9	2
10	2
10	2

To save the data that you entered:

> *File>Save as*
>> (Enter drive and file name for the save file.)
>> OK

Calculate the table and test statistic for the analysis:

> *Statistics>Nonparametric tests>Mann-Whitney*
>> Data variable: Behavior   [click to transfer]
>> Group var. (2 groups): Program
>> OK

Referring to the output, the rank sums for the two program groups are given as 65.5 and 105.5. The program uses another method of determining the probability, so U is not reported. The probability value reported at the bottom of the output is the two-tailed probability adjusted for ties; it may be slightly different from that obtained in the text (which does not adjust for ties).

   Clear the spreadsheet to input data for the next example:

> *File>New*
>> OK

StataQuest Program Output

```
. ranksum Behavior, by(Program)

Two-sample Wilcoxon rank-sum (Mann-Whitney) test

 Program |      obs     rank sum     expected
---------+---------------------------------
       1 |       10        65.5           95
       2 |        8       105.5           76
---------+---------------------------------
combined |       18         171          171

unadjusted variance        126.67
adjustment for ties         -2.61
                          ----------
adjusted variance          124.05

Ho: median Behavior(Program==1) = median Behavior(Program==2)
            z =    -2.649
    Prob > |z| =    0.0081
```

Kruskal-Wallis Test for *k* Independent Samples (Table 14-3)

The Kruskal-Wallis Test for *k* Independent Samples follows the same procedure as the Mann-Whitney, but it is used for three or more independent samples. In the example in Table 14-3, there are three program types coded: 1=Prosocial, 2=Neutral, 3=Violent.

Enter the values for *Behavior* in the first column of the Stata Editor and the codes for *Program* in the second column. Label the columns *Behavior* and *Program*. The spreadsheet data should appear as follows:

Behavior	Program
7	1
3	1
8	1
5	1
6	1
9	2
2	2
4	2
4	2
7	2
7	2
3	3
8	3
8	3
5	3
6	3
1	3

To save the data that you entered:

> *File>Save as*
> > (Enter drive and file name for the save file.)
> > OK

Calculate the chi-square test statistic for the analysis:

> *Statistics>Nonparametric Tests>Kruskal-Wallis*
> > Data variable: Behavior   [click to transfer]
> > Group variable: Program
> > OK

On the output, the chi-square statistic, degrees of freedom, and associated probability value are reported. Only results not corrected for ties are given by StataQuest, which corresponds to the method used in the text.

To clear the spreadsheet to input data for the next example:

> *File>New*
> > OK

StataQuest Program Output

```
. kwallis Behavior, by(Program)

Test: Equality of populations (Kruskal-Wallis Test)

  Program         _Obs      _RankSum
     1             5          47.50
     2             6          54.00
     3             6          51.50

chi-squared =      0.090 with 2 d.f.
probability =      0.9561
```

Wilcoxon Test for Two Correlated Samples (Table 14-4)

To calculate the Wilcoxon Test for correlated samples, the two measures must appear on a separate line for each subject, and the values for the two measures are entered in the first two columns of the Stata Editor spreadsheet. For the example in Table 14-4, label the two columns *Before* and *After*. The spreadsheet should appear as follows:

Before	After
4.5	5.2
3.0	4.5
6.2	5.9
4.9	5.2
5.5	5.8
5.8	6.1
6.1	7.2
7.0	7.1
5.0	5.6

To save the data that you entered:

> *File>Save as*
> > (Enter drive and file name for the save file.)
> > OK

Calculate the table and test statistic for the analysis:

> *Statistics>Nonparametric Tests>Wilcoxon signed-ranks*
> > Data variable #1: Before   [click to transfer]
> > Data variable #2: After
> > OK

The program uses another method than your text to rank differences and calculate the probability, so the sum of ranks may be different than your text and the output does not include a value for W_{obs}. Instead the program estimates a z value and reports the nondirectional probability associated with that z.

Clear the spreadsheet to input data for the next example:

> *File>New*
> > OK

StataQuest Program Output

```
. signrank Before = After

Wilcoxon signed-rank test

    sign |        obs   sum ranks      expected
---------+-----------------------------------
positive |          1           3         22.5
negative |          8          42         22.5
    zero |          0           0            0
---------+-----------------------------------
     all |          9          45           45

unadjusted variance          71.25
adjustment for ties          -0.50
adjustment for zeros          0.00
                           ----------
adjusted variance            70.75

Ho: median of Before = After
           z =   -2.318
    Prob > |z| =    0.0204
```

Spearman Rank-Order Correlation (Table 14-5)

The Spearman coefficient is equivalent to the Pearson correlation coefficient applied to ranks. Therefore it is a nonparametric measure of association to be used when the data are ordinal in nature. StataQuest will calculate the ranks and carry out the calculation directly using the Spearman command.

Enter the values of the Before and After aggressiveness ratings from Table 14-5 into the first two columns of the Stata Editor. Label the columns accordingly. The spreadsheet should appear as follows:

Before	After
6	−1
2	−3
3	8
8	10
9	17
1	7
4	5
7	16
10	14
5	8

To save the data that you entered:

> *File>Save as*
>> (Enter drive and file name for the save file.)
>> OK

Calculate the nonparametric correlation coefficient, Spearman's *rho*, with its associated level of significance:

> *Statistics>Correlation>Spearman (rank)*
>> Data variable #1: Before   [click to transfer]
>> Data variable #2: After
>> OK

The output gives the number of observations, the correlation *rho*, and an associated *p* value for the significance of the correlation.

> Print the log file for all five of the above analyses, and exit StataQuest:

> *File>Print Log*
>> OK

> *File>Exit*

StataQuest Program Output

```
. spearman Before After

 Number of obs =       10
Spearman's rho =        0.7356

Test of Ho: Before and After independent
      Pr > |t|  =        0.0153
```

Minitab

Chi-Square Test (Table 14-1)

Minitab will compute expected frequencies, the observed chi-square, and its associated probability value directly from an $r \times c$ contingency table. The data for this example are given in Table 14-1. There are three rows and five columns in the table of observed frequencies.

Start the Minitab program. In the Data window enter the observed frequencies directly into the first five columns of the spreadsheet. Do not label the column headings. The frequencies you have entered should appear on the spreadsheet as follows:

C1	C2	C3	C4	C5
60	20	130	110	30
40	70	115	200	25
10	40	55	60	35

To save the data that you entered:

> *File>Save Worksheet As*
> > (Enter drive and file name for the save file.)
> > OK

The chi square analysis may be carried out using the Chisquare Test command. This command will process a contingency table of observed frequencies up to seven columns wide. (If you have raw data and need to form the contingency table, use the Cross Tabulation command instead.)

> *Stat>Tables>Chisquare Test*
> > Columns containing the table: C1-C5   [highlight all five columns]
> > Select
> > OK

On the output for this analysis, the expected frequencies are printed below the observed frequencies for each cell. Note that the observed chi-square value (88.182) is so large that the p value, at the bottom, is listed as 0.000. Therefore the probability that such a value should occur by chance is $p < .001$.

Clear the spreadsheet to input data for the next example:

> *File>New Worksheet*

Minitab Program Output

Chi-Square Test

Expected counts are printed below observed counts

	C1	C2	C3	C4	C5	Total
1	60	20	130	110	30	350
	38.50	45.50	105.00	129.50	31.50	
2	40	70	115	200	25	450
	49.50	58.50	135.00	166.50	40.50	
3	10	40	55	60	35	200
	22.00	26.00	60.00	74.00	18.00	
Total	110	130	300	370	90	1000

Chi-Sq = 12.006 + 14.291 + 5.952 + 2.936 + 0.071 +
         1.823 + 2.261 + 2.963 + 6.740 + 5.932 +
         6.545 + 7.538 + 0.417 + 2.649 + 16.056 = 88.182

DF = 8, P-Value = 0.000

Mann-Whitney *U* Test (Table 14-2)

Minitab will calculate a Mann-Whitney *U* Test for two independent samples, but it does not report the observed *U*. However, it does give the probability of the observed *U* with and without adjustment for tied ranks. In the example in Table 14-2, observers rated children's positive behavior after watching one of two different television program types, violent or prosocial. The dependent variable is the behavior rating. There are two program groups to be compared, Prosocial and Violent.

For this analysis, Minitab requires that the data be entered in "unstacked" form, with the data for the two program groups in separate columns. In the Data window spreadsheet, enter the behavior ratings for the two program groups in the first two columns of the spreadsheet. Label the two columns *Psocial* and *Violent*. The data in the spreadsheet should now appear as follows:

Psocial	Violent
1	5
2	6
2	8
4	9
5	9
5	9
6	10
6	10
7	
9	

To save the data that you entered:

> *File>Save Worksheet As*
>> (Enter drive and file name for the save file.)
>> OK

To carry out the analysis which will calculate a table and test statistic for the sample:

> *Stat>Nonparametrics>Mann-Whitney*
>> First Sample: Psocial   [double-click to transfer]
>> Second Sample: Violent
>> Alternative: not equal
>> OK

In the output for this analysis, $W = 65.5$ is the sum of the ranks for *Psocial* which we call T_A (It is *not* U_{obs}, which must be calculated by hand using the formula in Table 14.2) The next-to-last line of the output gives the unadjusted probability (which is used in this text), and the last line adjusts this probability for ties (the adjusted value is not given in the Student Version of Minitab.) These values differ slightly from those used in the other programs because of the way the correction is made.

Clear the spreadsheet to input data for the next example:

> *File>New Worksheet*

Minitab Program Output

Mann-Whitney Confidence Interval and Test

```
Psocial    N =   10      Median =         5.000
Violent    N =    8      Median =         9.000
Point estimate for ETA1-ETA2 is       -4.000
95.4 Percent CI for ETA1-ETA2 is (-6.000,-0.999)
W = 65.5
Test of ETA1 = ETA2  vs  ETA1 not = ETA2 is significant at 0.0100
The test is significant at 0.0092 (adjusted for ties)
```

Kruskal-Wallis Test for *k* Independent Samples (Table 14-3)

The Kruskal-Wallis test is used when the number of groups being compared is greater than two. Minitab will compute a Kruskal-Wallis Test for *k* independent samples and requires that the data be entered using the "stacked" method. For each subject in Table 14-3, values of *Behavior*, the dependent or response variable, must be entered, followed by the kind of TV programs they viewed (1=Prosocial, 2=Neutral, 3=Violent).

In the Data window, enter the values for *Behavior* in the first column of the spreadsheet, and the *Program* code in the second column. Label the two columns *Behavior* and *Program*.

The spreadsheet data should appear as follows:

Behavior	Program
7	1
3	1
8	1
5	1
6	1
9	2
2	2
4	2
4	2
7	2
7	2
3	3
8	3
8	3
5	3
6	3
1	3

To save the data that you entered:

> *File>Save Worksheet As*
> > (Enter drive and file name for the save file.)
> > OK

Calculate the test statistic for the Kruskal-Wallis analysis:

> *Stat>Nonparametrics>Kruskal-Wallis*
> > Response: Behavior   [double-click to transfer]
> > Factor: Program   [double-click to transfer]
> > OK

The output reports the value of *H*, *df*, and *p* without (which is used in this text) and with adjustment for ties. Note that the Student Version of Minitab gives only the *H* adjusted for ties (but no probability level), so you must use the *H* with Table G in your text to obtain the probability level.

To clear the spreadsheet to input data for the next example:

File>New Worksheet

Minitab Program Output

Kruskal-Wallis Test

```
Kruskal-Wallis Test on Behavior

Program      N     Median    Ave Rank          Z
1            5      6.000         9.5       0.26
2            6      5.500         9.0       0.00
3            6      5.500         8.6      -0.25
Overall     17                   9.0

H = 0.09   DF = 2   P = 0.956
H = 0.09   DF = 2   P = 0.955 (adjusted for ties)
```

Wilcoxon Test for Two Correlated Samples (Table 14-4)

Minitab can be used to carry out the Wilcoxon test on the difference between two correlated (for example, before and after values) samples. Data must be entered as paired variables on the spreadsheet and a difference score calculated.

Enter the values from Table 14-4 for the two measures, *Before* and *After*, in the first two columns of the Data window spreadsheet, and label the columns accordingly. A difference score (*Diff*) must be calculated by subtracting the value of *Before* from the value of *After*. Define a new variable *Diff* by typing its name at the top of the third column of the spreadsheet. Then use the Calculator command to fill in the values of *Diff*:

> *Calc>Calculator*
> > Store result in variable: Diff   [double-click to transfer]
> > Expression: **After - Before**   [double-click, or type in]
> > OK

The spreadsheet, with the new variable, should appear as follows:

Before	After	Diff
4.5	5.2	0.7
3.0	4.5	1.5
6.2	5.9	−0.3
4.9	5.2	0.3
5.5	5.8	0.3
5.8	6.1	0.3
6.1	7.2	1.1
7.0	7.1	0.1
5.0	5.6	0.6

To save the data in the spreadsheet:

> *File>Save Worksheet As*
> > (Enter drive and file name for the save file.)
> > OK

Calculate the table and test statistic for the Wilcoxon Signed Rank test on the variable *Diff*:

> *Stat>Nonparametrics>1-Sample Wilcoxon*
> > Variables: Diff   [double-click to transfer]
> > Test median: Yes   Value: 0.0
> > Alternative: not equal
> > OK

The output reports N, the number of pairs, and the Wilcoxon statistic, in this case T_+, and its associated p value. Notice that the "Wilcoxon Statistic" on the output may or may not be the smaller of T_+ and T_- (in this case it is *not* the smaller value) which is taken to be W_{obs} in the text and used to enter Table I in the textbook to determine the probability. The probability on the output, however, should be accurate.

Clear the spreadsheet to input data for the next example:

File>New Worksheet

Minitab Program Output

Wilcoxon Signed Rank Test

Test of median = 0.000000 versus median not = 0.000000

	N	N for Test	Wilcoxon Statistic	P	Estimated Median
Diff	9	9	41.5	0.028	0.4500

Spearman Rank-Order Correlation (Table 14-5)

The Spearman rank-order correlation coefficient is equivalent to applying the Pearson correlation to ranked data. It is a nonparametric measure of association to be used when the data are ordinal in nature. Minitab does not calculate directly a Spearman rank-order correlation, but you can get the program to compute it by entering the data, ranking them, and then computing a Pearson correlation on the ranks.

Enter the pairs of aggressiveness ratings from Table 14-5 into the first two columns of the Minitab Data window. Label the two columns *Before* and *After*. Define two new variables by typing the names *Brank* and *Arank* at the top of the next two columns in preparation for determining rankings. Rankings are then stored in columns three and four using the following:

Manip>Rank
        Rank data in: Before   [double-click to transfer]
        Store ranks in: Brank
        OK
Manip>Rank
        Rank data in: After   [double-click to transfer]
        Store ranks in: Arank
        OK

The spreadsheet, with the rankings for the variables, should appear as follows:

Before	After	Brank	Arank
6	−1	6	2.0
2	−3	2	1.0
3	8	3	5.5
8	10	8	7.0
9	17	9	10.0
1	7	1	4.0
4	5	4	3.0
7	16	7	9.0
10	14	10	8.0
5	8	5	5.5

To save the raw data and the rankings as they appear on the spreadsheet above:

File>Save Worksheet As
        (Enter drive and file name for the save file.)
        OK

Finally, the correlation coefficient, Spearman's *rho*, is obtained by correlating the two rankings.

Stat>Basic Statistics>Correlation
        Variables: Brank Arank   [double-click to transfer]
        OK

The output consists of the Pearson correlation between the two rankings, which is the Spearman correlation. Minitab does not provide a level of significance for the correlation coefficient; use Table J in your text.
    To obtain the results for all five of the nonparametric analyses, print out the Session window:

File>Print Worksheet
        Print Range: All
        OK

Exit Minitab:

File>Exit

Minitab Program Output

Correlations (Pearson)

```
Correlation of Brank and Arank = 0.736
```

SPSS

Chi-Square Test (Table 14-1)

To compute the Pearson chi-square for a row-by-column set of independent frequencies such as in Table 14-1, SPSS requires three input variables: one that contains the frequencies for each cell, one for the row number, and one for the column number. In this example, there are three rows and five columns in the table of frequencies.

Start the SPSS program to bring up the Data Editor window. Enter the frequency data from Table 14-1 for each of the 15 cells, one cell at a time. The cell frequency, *Freq*, goes in the first variable on the spreadsheet. The row number, *Row*, goes in the second variable, and the column number, *Col*, goes in the third variable. Label the columns with the variable names *Freq*, *Row*, and *Col*. The data you have entered should appear on the spreadsheet as follows:

Freq	Row	Col
60	1	1
20	1	2
130	1	3
110	1	4
30	1	5
40	2	1
70	2	2
115	2	3
200	2	4
25	2	5
10	3	1
40	3	2
55	3	3
60	3	4
35	3	5

SPSS requires the Weight Cases command to multiply each cell by its frequency, so in effect it is reading information for 1000 cases (the total number of persons sampled) rather than the 15 cells that are entered on the spreadsheet. The Crosstabs command sets up the contingency table and calculates the chi-square statistic to test for independence between the table variables.

Weight the cases by cell frequency:

> *Data>Weight Cases*
> > Weight cases by: Yes
> > Frequency Variable: Freq   [highlight and transfer]
> > OK

Set up the contingency table and calculate the chi-square statistic:

> *Statistics>Summarize>Crosstabs*
>> Row(s): Row   [highlight and transfer]
>> Column(s): Col
>> Statistics:
>>> Chi-square: Yes
>>> Continue
>> Cells:
>>> Counts: Observed; Expected
>>> Percentages: None (Row, Column or Totals can be specified.)
>>> Continue
>> OK

The Crosstabulation table contains the frequencies (Counts) and the expected frequencies in each cell, and the Pearson Chi-square value and associated probability value appear on the first line of the Chi-square Tests table.

To print the output:

> *File>Print*
>> Print range: All visible output.
>> OK

To clear the output for the next part of the exercise:

> *File>Close*

To save the data that you entered and clear the spreadsheet for the next analysis:

> *Window>SPSS Data Editor*
> *File>Save as*
>> (Enter device and file name for SPSS save file.)
>> Save
> *File>New>Data*

SPSS Program Output

Crosstabs

Case Processing Summary

	Cases					
	Valid		Missing		Total	
	N	Percent	N	Percent	N	Percent
ROW * COL	1000	100.0%	0	.0%	1000	100.0%

ROW * COL Crosstabulation

			COL					Total
			1.00	2.00	3.00	4.00	5.00	
ROW	1.00	Count	60	20	130	110	30	350
		Expected Count	38.5	45.5	105.0	129.5	31.5	350.0
	2.00	Count	40	70	115	200	25	450
		Expected Count	49.5	58.5	135.0	166.5	40.5	450.0
	3.00	Count	10	40	55	60	35	200
		Expected Count	22.0	26.0	60.0	74.0	18.0	200.0
Total		Count	110	130	300	370	90	1000
		Expected Count	110.0	130.0	300.0	370.0	90.0	1000.0

Chi-Square Tests

	Value	df	Asymp. Sig. (2-tailed)
Pearson Chi-Square	88.182[a]	8	.000
Likelihood Ratio	87.972	8	.000
Linear-by-Linear Association	7.365	1	.007
N of Valid Cases	1000		

a. 0 cells (.0%) have expected count less than 5. The minimum expected count is 18.00.

Mann-Whitney *U* Test (Table 14-2)

The Mann-Whitney *U* Test compares two independent samples. In the case of Table 14-2, observers rated children's positive behavior after watching one of two different television programs, violent or prosocial. The dependent or test variable is the behavior rating, *Behavior*; the coded treatment or grouping variable is the program type, *Program* (1=Prosocial; 2=Violent).

The SPSS Data Editor should fill the screen. Enter the values from Table 14-2 for the behavior ratings in the first column and the code for the type of programming in the second column. Label the two columns with the variable names *Behavior* and *Program*. The data in the spreadsheet should now appear as follows:

Behavior	Program
1	1
2	1
2	1
4	1
5	1
5	1
6	1
6	1
7	1
9	1
5	2
6	2
8	2
9	2
9	2
9	2
10	2
10	2

Calculate the test statistics and tables for the analysis:

> *Statistics>Nonparametric Tests>2 Independent Samples*
> > Test Variable List: Behavior   [highlight and transfer]
> > Grouping Variable: Program (? ?)
> > Define Groups:
> > > Group 1: 1
> > > Group 2: 2
> > Test Type: Mann-Whitney U
> > > Continue
> > OK

The value of *U* reported by the program is 10.5. This may or may not be the one you obtain by other methods because there are two possible values of *U*, 10.5 and 69.5, depending on which group is designated *A* for the formula given in the text. The asymptotic probability corrected for ties is reported in the second to last line of the output, followed by the exact probability for a non-directional (2-tailed) test, which is not corrected for ties and is the one used in this text.

To print the output:

> *File>Print*
>> Print range: All visible output.
>> OK

To clear the output for the next part of the exercise:

> *File>Close*

To save the data that you entered:

> *Window>SPSS Data Editor*
> *File>Save as*
>> (Enter device and file name for SPSS save file.)
>> Save

SPSS Program Output

Mann-Whitney Test

Ranks

	PROGRAM	N	Mean Rank	Sum of Ranks
BEHAVIOR	1.00	10	6.55	65.50
	2.00	8	13.19	105.50
	Total	18		

Test Statistics[a]

	BEHAVIOR
Mann-Whitney U	10.500
Wilcoxon W	65.500
Z	-2.649
Asymp. Sig. (2-tailed)	.008
Exact Sig. [2*(1-tailed Sig.)]	.006[b]

a. Grouping Variable: PROGRAM

b. Not corrected for ties.

Kruskal-Wallis Test for *k* Independent Samples (Table 14-3)

The Kruskal-Wallis Test for *k* Independent Samples follows the same procedure as the Mann-Whitney but is for three or more independent samples. In the example in Table 14-3, there are three program types, so the coding for the variable *Program* ranges from 1 to 3.

Enter the values for *Behavior* from Table 14-3 in the first column of the SPSS Data Editor and the codes for *Program* (1=Prosocial, 2=Neutral, 3=Violent) in the second column. The spreadsheet data should appear as follows:

Behavior	Program
7	1
3	1
8	1
5	1
6	1
9	2
2	2
4	2
4	2
7	2
7	2
3	3
8	3
8	3
5	3
6	3
1	3

Calculate the test statistics and tables for the analysis:

> *Statistics>Nonparametric Tests>k Independent Samples*
> > Test Variable List: Behavior   [highlight and transfer]
> > Grouping Variable: Program (? ?)
> > Define Range:
> > > Minimum: 1
> > > Maximum: 3
> > Test Type: Kruskal-Wallis H
> > > Continue
> > OK

On the output, the chi-square statistic, degrees of freedom (df), and associated probability value (Asymp. Sig.) are reported. Only results corrected for ties are given, which may be different from your text which does not make this correction.

To print the output:

> *File>Print*
> > Print range: All visible output.
> > OK

To clear the output for the next part of the exercise:

File>Close

To save the data that you entered and clear the spreadsheet for the next analysis:

Window>SPSS Data Editor
File>Save as
        (Enter device and file name for SPSS save file.)
        Save
File>New>Data

SPSS Program Output

NPar Test

Kruskal-Wallis Test

Ranks

	PROGRAM	N	Mean Rank
BEHAVIOR 1.00		5	9.50
	2.00	6	9.00
	3.00	6	8.58
	Total	17	

Test Statistics[a,b]

	BEHAVIOR
Chi-Square	.091
df	2
Asymp. Sig.	.955

a. Kruskal Wallis Test

b. Grouping Variable: PROGRAM

Wilcoxon Test for Two Correlated Samples (Table 14-4)

To calculate a Wilcoxon Test for two correlated samples, two paired measures are entered on a separate line for each subject.

Call up the SPSS Data Editor spreadsheet and enter the values for the two measures, *Before* and *After*, in the first two columns of the spreadsheet. Label the columns accordingly. The spreadsheet should appear as follows:

Before	After
4.5	5.2
3.0	4.5
6.2	5.9
4.9	5.2
5.5	5.8
5.8	6.1
6.1	7.2
7.0	7.1
5.0	5.6

Calculate the tables and test statistics for the analysis:

> *Statistics>Nonparametric Tests>2 Related Samples*
>> Test Pair(s) List: After — Before   [highlight and transfer]
>> Test Type: Wilcoxon
>> OK

The program may rank the differences in a way different from your text, producing a different sum of ranks. Also, the Wilcoxon *W* statistic is not given on the output; only a z score and its associated *p* value (Asymp. Sig.) is reported.
   To print the output:

> *File>Print*
>> Print range: All visible output.
>> OK

To clear the output for the next part of the exercise:

> *File>Close*

To save the data that you entered:

> *Window>SPSS Data Editor*
> *File>Save as*
>> (Enter device and file name for SPSS save file.)
>> Save

SPSS Program Output

NPar Tests

Wilcoxon Signed Ranks Test

Ranks

		N	Mean Rank	Sum of Ranks
AFTER - BEFORE	Negative Ranks	8[a]	5.25	42.00
	Positive Ranks	1[b]	3.00	3.00
	Ties	0[c]		
	Total	9		

a. AFTER < BEFORE
b. AFTER > BEFORE
c. AFTER = BEFORE

Test Statistics[a]

	AFTER - BEFORE
Z	-2.318[b]
Asymp. Sig. (2-tailed)	.020

a. Wilcoxon Signed Ranks Test
b. Based on positive ranks.

Spearman Rank-Order Correlation (Table 14-5)

The Spearman coefficient is equivalent to the Pearson correlation coefficient applied to ranks. As a nonparametric measure of association, it is used when the data are ordinal. SPSS will calculate the ranks and carry out the calculation directly as part of the Bivariate command.

Enter the values of the before and after aggressiveness ratings from Table 14-5 into the first two columns of the SPSS Data Editor and label the columns *Before* and *After*. The spreadsheet should appear as follows:

Before	After
6	−1
2	−3
3	8
8	10
9	17
1	7
4	5
7	16
10	14
5	8

Use the following to calculate the value of the nonparametric correlation coefficient:

> *Statistics>Correlate>Bivariate*
>> Variables: After, Before   [highlight and transfer]
>> Correlation Coefficients: Spearman
>> Test of Significance: Two-tailed
>> OK

The output reports Spearman's *rho* and the associated two-tailed *p* value (Sig.).

To print the output:

> *File>Print*
>> Print range: All visible output.
>> OK

To save the data that you entered:

> *Window>SPSS Data Editor*
> *File>Save as*
>> (Enter device and file name for SPSS save file.)
>> Save

Exit SPSS:

> *File>Exit SPSS*

SPSS Program Output

Nonparametric Correlations

Correlations

			AFTER	BEFORE
Spearman's rho	Correlation Coefficient	AFTER	1.000	.736*
		BEFORE	.736*	1.000
	Sig. (2-tailed)	AFTER	.	.015
		BEFORE	.015	.
	N	AFTER	10	10
		BEFORE	10	10

*. Correlation is significant at the .05 level (2-tailed).

APPENDIX

REVIEW OF BASIC MATHEMATICS AND ALGEBRA

SYMBOLS AND SIGNS

Relationships between quantities can be expressed by symbols. If a and b are numerically identical, we write _____ . However, if a and b are not equal, this fact can be expressed by _____ . We use > to mean "is greater than" and < to mean "is less than." Thus 5 ___ 3 and 3 ___ 5. You can remember the difference between the two symbols by noting that the open end (larger end) of the symbols is always on the side of the larger number. Therefore, if a is greater than b, we write a ___ b or b ___ a.

    If one quantity is greater than or equal to another, we use a combination of the equality and greater than symbols, $\geq$. *If c is greater than or equal to d*, we write c ___ d or d ___ c. If t is a positive number or zero, we can write t ___ 0. If t is negative or zero, _____ . If t falls between 10 and 20, we can say that 10 is less than t, in symbols, _____ , and that t is less than 20, _____ . These two statements can be combined into $10 < t < 20$. If t can either equal 10 or 20 or fall somewhere between them, we write _____ .

    Some numbers are negative (i.e., below 0); that introduces the problem of signs when we perform elementary mathematical operations: $7 + (-4) =$ ___ , $7 - (-4) =$ ___ , and $-a - (-b) =$ _____ .

    The sign of the product obtained by multiplying several quantities together depends upon the number of negative values to be

$a = b$
$a \neq b$

$>; <$

$>; <$

$\geq; \leq$
$\geq$
$t \leq 0$
$10 < t\,; t < 20$

$10 \leq t \leq 20$

$3; 11; -a + b$

multiplied. If there is an even number of negative values among the
values to be multiplied, the product is positive. For example,
$(2)(4) =$ ___ , $(-2)(-4) =$ ___ , $(2)(-1)(-3) =$ ___ ,

$(1)(-2)(4)(-3) =$ ___ , $(-x)y(-z) =$ ___ . If there is an odd
number of negative values to be multiplied, the product is negative:
$(-1)(4) =$ ___ , $(3)(-2) =$ ___ , $(-2)(-3)(-1)(2) =$ ___ , and

$(a)(-b)(c) =$ ___ .

8; 8; 6

24; xyz

$-4; -6; -12$

$-abc$

The same rule applies to division. If there is an even number of
negative values, the result is positive: $6/2 =$ ___ , $-14/-7 =$ ___ ,
$(-2)(-3)/(-1)(-4) =$ ___ , and $-a/b(-c) =$ ___ . If there is an odd
number of negative values in the division, the result is negative:
$-8/2 =$ ___ , $12/-4 =$ ___ , $(-2)(4)/(3) =$ ___ , and

$(-a)(-b)/c(-d) =$ ___ .

3; 2

$3/2$; a/bc

$-4; -3; -8/3$

$-\dfrac{ab}{cd}$

At times we want to disregard the algebraic sign of a quantity
and consider only its absolute value. The absolute value of a is
written $|a|$. To disregard the sign of d we write _____ , and the
absolute value of $(X - Y)$ is symbolized _____ . For example,
$|-4| =$ ___ , and $|6| =$ ___ .

$|d|$

$|X - Y|$

4; 6

OPERATIONS

Fractions

The product of two or more fractions equals the product of the
numerators divided by the product of the denominators: $\dfrac{1}{2} \cdot \dfrac{3}{5} =$ ___ ,

$\dfrac{1}{3} \cdot \dfrac{1}{5} \cdot \dfrac{2}{7} =$ ___ , and $\dfrac{a}{c} \cdot \dfrac{b}{d} =$ ___ . To divide one fraction by another,
invert the divisor (the fraction you want to divide *by*) and multiply.
To divide $\dfrac{1}{2}$ by $\dfrac{3}{5}$, invert the divisor to obtain ___ and then multiply,

_____ $=$ ___ . Similarly, $\dfrac{x}{y} \div \dfrac{a}{b} =$ _____ $=$ ___ .

$\dfrac{3}{10}$

$\dfrac{2}{105}$; $\dfrac{ab}{cd}$

$\dfrac{5}{3}$

$\dfrac{1}{2} \cdot \dfrac{5}{3} = \dfrac{5}{6}$; $\dfrac{x}{y} \cdot \dfrac{b}{a} = \dfrac{xb}{ya}$

Fractions often can be simplified by **factoring** and **cancellation**.
Factoring involves determining the simplest set of numbers which,
when multiplied together, will yield the original number. The number
12 can be factored into $(3)(4)$, and the 4 can be further factored into
$(2)(2)$, yielding $(3)(2)(2)$ as the simplest factored equivalent of 12. In
the same way, 18 can be factored into _____ , 24 into
_____ , and 64 into _____ .

$(2)(3)(3)$

$(3)(2)(2)(2)$; $(2)(2)(2)(2)(2)(2)$

To simplify fractions, factor the numerator and the denominator
and cancel numbers that appear in both the numerator and

denominator: $\dfrac{6}{9} = \dfrac{(2)(\cancel{3})}{(3)(\cancel{3})} = \dfrac{2}{3}$. Thus $\dfrac{12}{36} = $ _____ = ___

$\dfrac{(\cancel{2})(\cancel{2})(\cancel{3})}{(\cancel{2})(\cancel{2})(\cancel{3})(3)} = \dfrac{1}{3}$

and $\dfrac{64}{72} = $ _____ = ___ .

$\dfrac{(\cancel{2})(\cancel{2})(\cancel{2})(2)(2)(2)}{(\cancel{2})(\cancel{2})(\cancel{2})(3)(3)} = \dfrac{8}{9}$

The multiplication of fractions can be simplified by factoring and canceling any terms that occur in both the numerators and denominators: $\left(\dfrac{2}{3}\right)\left(\dfrac{3}{4}\right) = $ _____ = _____ and

$\left(\dfrac{\cancel{2}}{\cancel{3}}\right)\left(\dfrac{\cancel{3}}{\cancel{2}\cdot 2}\right) = \dfrac{1}{2}$

$\left(\dfrac{3}{5}\right)\left(\dfrac{2}{3}\right)\left(\dfrac{5}{6}\right) = $ _____ = _____ .

$\left(\dfrac{\cancel{3}}{\cancel{5}}\right)\left(\dfrac{\cancel{2}}{\cancel{3}}\right)\left(\dfrac{\cancel{5}}{3\cdot\cancel{2}}\right) = \dfrac{1}{3}$

Two fractions can be added or subtracted only if they both have the same denominator. Therefore, fractions that do not have the same denominator must be converted to ones having the same denominator. This is done by multiplying the numerator and denominator of a fraction by the same number, often the denominator of the other fraction. To add 1/2 and 1/3, change them both into

fractions with denominators 6. The fraction $\dfrac{1}{2}$ can be changed by

multiplying it by $\dfrac{3}{3}$ to produce ___ , and $\dfrac{1}{3}$ can be changed by

$\dfrac{3}{6}$

multiplying it by $\dfrac{2}{2}$ to produce ___ . Then their numerators are

$\dfrac{2}{6}$

simply added and their sum divided by the common denominator:

$\dfrac{3}{6} + \dfrac{2}{6} = $ ___ . To add $\dfrac{a}{b}$ and $\dfrac{c}{d}$ convert both fractions to ones

$\dfrac{5}{6}$

having a common denominator _____ . Then add the

$\dfrac{ad}{bd} + \dfrac{cd}{bd}$

numerators and divide by the common denominator, _____ .

$\dfrac{ad+cd}{bd}$

Subtraction follows a similar process: $\dfrac{1}{4} - \dfrac{1}{5} = $ _____ = ___ .

$\dfrac{5}{20} - \dfrac{4}{20} = \dfrac{1}{20}$

Factorials

When you calculate certain formulas, it will be necessary to use factorials. For example, 3!, read "three factorial," is a simplified way of writing (3)(2)(1). In the same way, 5! = _____ = ___ .

$(5)(4)(3)(2)(1) = 120$

$(5-3)! = $ _____ = ___ , and

$2! = (2)(1) = 2$

$\dfrac{5!}{3!} = $ _____ = ___ . It is important to know that

$\dfrac{(5)(4)(\cancel{3})(\cancel{2})(\cancel{1})}{(\cancel{3})(\cancel{2})(\cancel{1})} = 20$

$0! = 1$; thus, $\dfrac{3!}{0!} = $ _____ .

$\dfrac{(3)(2)(1)}{(1)} = 6$

Exponents

Exponents, small numbers written after and above a base number, signify that the base number is to be multiplied by itself as many times as the exponent states. Thus, $3^2 = (3)(3) = 9$ and

$3^3 = $ _____ $= $ ___ . It is important to distinguish between

$(2+3+5)^2 = $ _____ $= $ _____

$(3)(3)(3) = 27$

$10^2 = 100$

and

$2^2 + 3^2 + 5^2 = $ _____ $= $ _____

$4 + 9 + 25 = 38$

For the following set of scores, then,

1
6
2
4

the sum is ___ , the squared sum is ____ , and the sum of squared scores is ___ .

$13; 169$
57

Note that $n^1 = n$ and that $n^0 = 1$.

Generally, numbers with exponents cannot be added except by carrying out the exponentiation and then adding. Thus,
$2^2 + 2^3 = (2)(2) + (2)(2)(2) = 4 + 8 = 12$ and

$2^4 - 3^2 = $ _____ $= $ _____ $= $ ___ .
However, the *product* of two exponential quantities *having the same base number* is equal to that same number raised to the *sum* of the exponents, as $(2^3)(2^4) = 2^{3+4} = 2^7$, $(3^2)(3^3) = $ _____ $= $ ___ ,

and $(r^s)(r^t) = $ _____ . To divide two exponential quantities *having the same base number,* subtract their exponents. For example,

$\dfrac{2^5}{2^4} = 2^{5-4} = 2$, $\dfrac{3^6}{3^2} = $ _____ $= $ ___ , and $\dfrac{r^s}{r^t} = $ _____ .

Remember: to perform these multiplication and division operations on exponential quantities, the two quantities must have the

_____ _____ _____ . A fraction like $\dfrac{2^2}{3^2}$, in which

the base numbers in the numerator and denominator are different, is simplified by carrying out the exponentiation separately within the numerator and denominator, _____ $= $ ___ .

$(2)(2)(2)(2) - (3)(3) = 16 - 9 = 7$

$3^{2+3} = 3^5$

r^{s+t}

$3^{6-2} = 3^4; r^{s-t}$

same base number

$\dfrac{(2)(2)}{(3)(3)} = \dfrac{4}{9}$

To raise a fraction to a power, raise the numerator and the

denominator to that same power. Thus, $\left(\dfrac{2}{3}\right)^2 = \left(\dfrac{2^2}{3^2}\right)$, $\left(\dfrac{3}{4}\right)^2 = $ ___ ,

$\dfrac{3^2}{4^2}$

and $\left(\dfrac{s}{t}\right)^r =$ _____ .

$\dfrac{s^r}{t^r}$

Binomial Expansion

It will sometimes be necessary to square a binomial, which is an expression involving the addition or subtraction of two quantities, such as $(a+b)^2$. The result is the square of the first term, plus two times the product of the two terms, plus the square of the second term. Thus, $(a+b)^2 =$ _____ . If a negative term is involved, the process is the same, but greater attention must be paid to the signs: $(a-b)^2 =$ _____ . This technique is quite general and can be applied to any binomial. For example,
$(X-\overline{X})^2 =$ _____ and
$98^2 = (100-2)^2 =$ _____ = ____ .

$a^2 + 2ab + b^2$

$a^2 - 2ab + b^2$

$X^2 - 2X\overline{X} + \overline{X}^2$
$10{,}000 - 400 + 4 = 9604$

Square Roots

Although most square roots can be determined on a hand calculator, it is helpful to know some simple algebra pertaining to square roots. The $\sqrt{25} =$ ____ and $\sqrt{.25} =$ ____ ,but $\sqrt{2.5}$ is 1.58. Notice that $\sqrt{4x} = \sqrt{4}\sqrt{x} =$ _____ , $\sqrt{16x^2} = \sqrt{16}\sqrt{x^2} =$ _____ , and $\sqrt{\dfrac{4}{11}} = \dfrac{\sqrt{4}}{\sqrt{11}} =$ _____ . On the other hand, $\sqrt{9+x^2}$ cannot be reduced further, and $\sqrt{a^2-b^2}$ does *not* equal $a-b$.

$5; .5$

$2\sqrt{x} ; 4x$

$\dfrac{2}{\sqrt{11}}$

Factoring and Simplification

As explained in the text, the process of breaking down a number into parts which, when multiplied together, equal the number is called _____ . The process of reducing the number of quantities in an expression is called _____ .

Occasionally, algebraic expressions may be simplified by removing parentheses. When no multiplication or division is involved, remove parentheses and perform any addition or subtraction necessary:
$4+(-2) = 4-2 = 2$, $-5-(-4)+(-1) =$ _____ = ____ , and
$(a)-(-b)-(c+d) =$ _____ . When several different operations are required, it is best to simplify the expression by following a specific sequence of operations. Suppose you are required to simplify

$$700 - 2(4+6)^2$$

factoring
simplification

$-5+4-1 = -2$
$a+b-c-d$

First, perform any additions or subtractions within terms or parentheses. Thus, the expression becomes

$$700 - 2\left(\underline{\hspace{1cm}}\right)^2 \qquad\qquad 10$$

Second, perform any exponentiations or square roots:

$$700 - 2\left(\underline{\hspace{1cm}}\right) \qquad\qquad 100$$

Third, calculate any multiplications or divisions:

$$700 - \underline{\hspace{2cm}} \qquad\qquad 200$$

Finally, perform any addition or subtraction between terms:

$$700 - 200 = \underline{\hspace{2cm}} \qquad\qquad 500$$

Following the same sequence, the expression

$$(4+2) - 2\sqrt{2} + 7$$

can be simplified by

first, $\left(\underline{\hspace{1cm}}\right) - 2\sqrt{\underline{\hspace{1cm}}}$ $\qquad$ 6; 9

second, $6 - 2\left(\underline{\hspace{1cm}}\right)$ $\qquad$ 3

third, $6 - \underline{\hspace{2cm}}$ $\qquad$ 6

and fourth, $= \underline{\hspace{2cm}}$ $\qquad$ 0

Sometimes it is helpful to transpose terms from one side of an expression to the other. The simplest method is to add (or subtract) the same term on both sides of the equation. For example, to get b on one side of the following equation, subtract a from each side:

$$a - b = c$$

$$\underline{}\quad -a = -a$$

$$a - b - a = c - a$$

$$-b = c - a$$

and to solve for X in $2X - 3 = 1$,

$$2X - 3 = 1 \qquad\qquad\qquad 3 + 3$$

$$\underline{\hspace{3cm}} \qquad\qquad\qquad 2X = 4$$

$$\underline{\hspace{3cm}}$$

This last expression can be solved for X by dividing each side of the equation by 2:

$$\frac{2X}{2} = \frac{4}{2}$$

$$X = \underline{\hspace{2cm}} \qquad\qquad\qquad 2$$

Similarly, if you want to solve for a in $\dfrac{a}{b} = c$, multiply both sides of

the equation by $\underline{\hspace{0.7cm}}$ giving $\underline{\hspace{2.5cm}}$. Then cancel, giving $b; \; b\left(\dfrac{a}{b}\right) = bc$

$$\underline{\hspace{2cm}} \;. \qquad\qquad\qquad a = bc$$

To *factor* an algebraic expression, break it down into a set of components that, when multiplied, equal the original expression. For example, $ab + ac = a(b + c)$; $ab - ac =$ _____ ;

$-ab - ac =$ _____ ; and

$ab + ac + db + dc =$ _____ = _____ .

$a(b - c)$

$-a(b + c)$

$a(b + c) + d(b + c) = (a + d)(b + c)$

Sometimes in the course of simplifying an expression it helps to divide each term in the numerator by the denominator:

$\dfrac{a - b}{c} = \dfrac{a}{c} - \dfrac{b}{c}$, and $\dfrac{ac - b}{b} =$ _____ = _____ . Be

$\dfrac{ac}{b} - \dfrac{b}{b} = \dfrac{ac}{b} - 1$

careful not to make the following mistake: $\dfrac{a}{b + c}$ is *not* equal to

$\dfrac{a}{b} + \dfrac{a}{c}$.

SELF-TEST AND EXERCISES

1. Using the less than and greater than symbols, express the unequal relationship between 2 and 4 in two ways.

2. State symbolically the fact that X:
   a. falls between 10 and 20
   b. is negative
   c. is greater than or equal to 0
   d. is less than or equal to 0 and greater than -10

3. Simplify
   a. $(-3)(-2)$
   b. $(-a)(b)(-c)$
   c. $-12/6$
   d. $-10 + 5$
   e. $5 - \dfrac{(-23)(-5)}{(-23)}$

4. Simplify and solve the following:
   a. $\dfrac{2}{3} \cdot \dfrac{1}{4}$
   b. $\dfrac{2}{3} \div \dfrac{1}{4}$
   c. $\dfrac{2}{3} + \dfrac{1}{4}$
   d. $\left(\dfrac{1}{2} + \dfrac{7}{10}\right)\left(\dfrac{2}{3} - \dfrac{1}{4}\right)$
   e. $6!/4!$

5. Simplify.
   a. $\dfrac{28}{32}$
   b. $\left(\dfrac{3}{5}\right)\left(\dfrac{2}{3}\right)\left(\dfrac{15}{16}\right)$

6. Simplify
   a. 3^4
   b. 2^{5-3}
   c. $\left(\dfrac{2}{3}\right)^4$
   d. $(a + b)^2$
   e. $(2 - b)^2$

7. Simplify.
   a. $5 - 3(2) + (3 + 4)$
   b. $5X - 10 = 20$
   c. $\dfrac{ab + bc}{b(a + c)}$
   d. $\dfrac{a(b - c) + (ab - ac)}{-a(c - b)}$
   e. $x(y + z) = xy + 2$
   f. $\dfrac{-1(-ab) - ac + a^2}{a} + c$

8. Determine the sum, the square of the sum, and the sum of the squared scores for:

   5
   -3
   2
   2
   -5
   1

ANSWERS

(1) $2 < 4$, $4 > 2$. **(2a)** $10 < X < 20$; **(2b)** $X < 0$; **(2c)** $X \geq 0$; **(2d)** $-10 < X \leq 0$. **(3a)** 6; **(3b)** abc; **(3c)** -2; **(3d)** -5; **(3e)** 10. **(4a)** $\frac{1}{6}$; **(4b)** $\frac{8}{3}$; **(4c)** $\frac{11}{12}$; **(4d)** $\frac{1}{2}$; **(4e)** 30. **(5a)** $\frac{7}{8}$; **(5b)** $\frac{3}{8}$.

(6a) 81; **(6b)** 4; **(6c)** $\frac{16}{81}$; **(6d)** $a^2 + 2ab + b^2$; **(6e)** $4 - 4b + b^2$. **(7a)** 6; **(7b)** $X = 6$; **(7c)** 1; **(7d)** 2; **(7e)** $xz = 2$; **(7f)** $a + b$. **(8)** 2, 4, 68.